Springer Theses

Recognizing Outstanding Ph.D. Research

Aims and Scope

The series "Springer Theses" brings together a selection of the very best Ph.D. theses from around the world and across the physical sciences. Nominated and endorsed by two recognized specialists, each published volume has been selected for its scientific excellence and the high impact of its contents for the pertinent field of research. For greater accessibility to non-specialists, the published versions include an extended introduction, as well as a foreword by the student's supervisor explaining the special relevance of the work for the field. As a whole, the series will provide a valuable resource both for newcomers to the research fields described, and for other scientists seeking detailed background information on special questions. Finally, it provides an accredited documentation of the valuable contributions made by today's younger generation of scientists.

Theses may be nominated for publication in this series by heads of department at internationally leading universities or institutes and should fulfill all of the following criteria

- They must be written in good English.
- The topic should fall within the confines of Chemistry, Physics, Earth Sciences, Engineering and related interdisciplinary fields such as Materials, Nanoscience, Chemical Engineering, Complex Systems and Biophysics.
- The work reported in the thesis must represent a significant scientific advance.
- If the thesis includes previously published material, permission to reproduce this must be gained from the respective copyright holder (a maximum 30% of the thesis should be a verbatim reproduction from the author's previous publications).
- They must have been examined and passed during the 12 months prior to nomination.
- Each thesis should include a foreword by the supervisor outlining the significance of its content.
- The theses should have a clearly defined structure including an introduction accessible to new PhD students and scientists not expert in the relevant field.

Indexed by zbMATH.

More information about this series at https://link.springer.com/bookseries/8790

Takayuki Kurihara

Observation and Control of Magnetic Order Dynamics by Terahertz Magnetic Nearfield

Doctoral Thesis accepted by
The University of Tokyo, Tokyo, Japan

Springer

Author
Dr. Takayuki Kurihara
Institute for Solid State Physics
The University of Tokyo
Kashiwa, Chiba, Japan

Supervisor
Prof. Tohru Suemoto
Emeritus Professor of The University of Tokyo
Tokyo, Japan

ISSN 2190-5053 ISSN 2190-5061 (electronic)
Springer Theses
ISBN 978-981-16-8792-1 ISBN 978-981-16-8793-8 (eBook)
https://doi.org/10.1007/978-981-16-8793-8

This Springer imprint is published by the registered company Springer Nature Singapore Pte Ltd.
The registered company address is: 152 Beach Road, #21-01/04 Gateway East, Singapore 189721, Singapore

Supervisor's Foreword

Control of ultrafast spin dynamics in the terahertz (THz) frequency region is a cutting-edge technology that is expected to enable picosecond-fast operation of spintronics devices. The magnetic field component of the THz pulse can drive spin dynamics through Zeeman interaction, without causing higher-energy electronic transitions. The possibility of manipulating *macroscopic* magnetic order with intense THz waves has thus attracted interest from both viewpoints of the fundamental scientific and spintronics application. Despite a number of theoretical predictions, however, the control of magnetic order at a macroscopic level with THz magnetic field has not been realized up to this point.

In order to realize such large-amplitude manipulation of spins using THz magnetic fields, in this Thesis Kurihara focused on a spin system which responds highly sensitively to the THz magnetic fields, that is, the spin reorientation phase transition (SRPT) that occurs in erbium orthoferrite ($ErFeO_3$). In addition, to increase the available strength of the THz magnetic field, he introduced a magnetic plasmonic structure called split-ring resonator (SRR). In this thesis, this novel combination of plasmonics and magnetic phase transition allowed a very clear-cut demonstration of the coherent control of the macroscopic magnetization: undoubtedly a milestone in the field of ultrafast magnonics.

In Chap. 1, a thorough introductory review covering the related research fields, namely THz spectroscopy, plasmonics and ultrafast spintronics, is given. In Chap. 2, further descriptions of the experimental sample material, orthoferrite $ErFeO_3$, the technique of intense THz pulse generation and THz metamaterial are described in more detail.

Chapter 3 consists of the first experimental results. Here, the basic characteristics of the spin dynamics resonantly excited by the SRR structure is presented. It was shown that the THz magnetic nearfield around the SRR structure is capable of enhancing the spin precession, by a factor reaching nearly an order of magnitude compared to direct impulsive excitation by incident THz field. Detailed analysis showed that the lifetime of spin precession excited by the SRR magnetic field is shortened compared to the original lifetime. This indicates the existence of interactive energy transfer between spin system and SRR mediated by magnetic nearfields, which caused the energy of the spin system to be consumed by Joule loss in the SRR structure.

Chapter 4 includes the most important discovery of the thesis. Here, plasmonically enhanced THz magnetic nearfield induces a coherent spin presession at the initial stage of the ultrafast optical heat-assisted SRPT. As a result, the direction of magnetization after SRPT has been shown to align almost uniformly, breaking the symmetry of the otherwise degenerated phase transition paths. It is revealed that the observed magnetization alignment during SRPT is crucially determined by the instantaneous phase of the coherent spin precession at the moment of SRPT when the coherently induced spin precession amplitude is large enough.

In Chap. 5, the case where the precession amplitude is small is presented, followed by a detailed spin dynamics calculation using Landau–Lifshitz–Gilbert equations. By using realistic parameters for the antiferromagnetic spin Hamiltonian, it is clearly revealed that the shift of the free energy caused by the transient THz magnetic field becomes the dominant mechanism for the magnetization symmetry breaking.

In summary, Kurihara's thesis presents unique and important experimental results of macroscopic magnetic control by THz pulses that combines plasmonics and magnetic phase transition, which I believe is interesting for the scholars in condensed matter and ultrafast photonics.

Tokyo, Japan
September 2021

Prof. Tohru Suemoto

Parts of this thesis have been published in the following journal articles:

Peer-Reviewed Papers

T. Kurihara, H. Watanabe, M. Nakajima, S. Karube, K. Oto, Y. Otani and T. Suemoto, "Macroscopic Magnetization Control by Symmetry Breaking of Photoinduced Spin Reorientation with Intense Terahertz Magnetic Near Field", Phys. Rev. Lett. **120**, 107202 (2018)

T. Kurihara, K. Nakamura, K. Yamaguchi, Y. Sekine, Y. Saito, M. Nakajima, K. Oto, H. Watanabe and T. Suemoto, "Enhanced spin-precession dynamics in a spin-metamaterial coupled resonator observed in terahertz time-domain measurements", Physical Review B **90**, 144408 (2014).

Conference Contributions

T. Kurihara, H. Watanabe, Y. Sekine, Y. Saito, M. Nakajima, K. Oto and T. Suemoto "Terahertz magnetic control of spin orientation during the laser-induced spin reorientation transition in orthoferrites using split-ring resonator", FTT 2015, Hamamatsu,Japan, 2015.9.1.

T. Kurihara, K. Nakamura, K. Yamaguchi, Y. Sekine, Y. Saito, M. Nakajima, K. Oto, H. Watanabe, and T. Suemoto, "Interactive Magnetic Coupling Between Spin Precession and Split-Ring Resonator in the Terahertz Frequency", IRMMW-THz 2014, University of Arizona (Tucson, USA), 28 Sept. 2014.

Acknowledgements

I am deeply thankful to Prof. T. Suemoto for his instructions and enormous support during the past 5 years including my master and Ph.D. course.

I would also like to thank Dr. M. Nakajima (Associate Professor at Osaka University, Research Associate at Suemoto group until March 2012) and Dr. H. Watanabe (Research Associate at Osaka University, Research Associate at S. lab until September 2015) for their supports and useful advices throughout the research.

I would also like to acknowledge external collaborators. Prof. Oto (Chiba University) and Prof. Otani (Nanoscale Science Division, ISSP) supported on the fabrication of SRR structures used in the study. Mr. Kitazawa (Materials Synthesis Section, ISSP) gave me instructions and support on the crystal growth of orthoferrite. Special thanks goes to Prof. Kato (ISSP) for the insightful discussions and advices on the theoretical treatments of spin dynamics. Professor Keith Nelson (MIT) and his group members gave me an opportunity to visit his group for collaboration during my 2nd year of Ph.D.

Additionally, I would also like to show my appreciations to my colleagues and alumni of Suemoto group: Dr. J. Nishitani, Dr. Y. Minami, Dr. K. Yamaguchi, Dr. A. Akifumi, Mr. K. Nakamura, Mr. M. Takeda, Mr. M. Hashimoto, Mr. T. Kawasaki, Mr. S. Ishige, Mr. R. Takei, Mr. S. Sakaki and Mr. S. Y. Maezawa. It was a great experience to work together and discuss physics day and night with them.

Furthermore, I am grateful for my dissertation committee for reviewing my thesis: Prof. H. Akiyama, Prof. S. Takeyama, Prof. T. Kato, Prof. R. Shimano and Prof. N. Kida.

Finally, I would like to thank my parents for their constant and heartful supports.

Contents

Chapter 1
Introduction

Abstract As an introductory chapter, a brief overview of the research field relevant for the thesis is summarized. The topics include terahertz spectroscopy, ultrafast magnonics, and plasmonics.

Keywords Terahertz spectroscopy · Plasmonics · Magnetic field · Spintronics · Spin reorientation

1.1 Terahertz (THz) Control of Magnetic System

1.1.1 THz Spectroscopy

Terahertz (THz) is a term used to indicate the frequency of 10^{12} Hz. This frequency region (around 0.1–10 THz, or wavelength of several tens of μm to mm) lies above the limit of operation frequency of electronics technology (~10 GHz). It is also much below the frequencies where conventional optical spectroscopic techniques are applicable, because the photon energy of THz waves is below that of background thermal radiation of room-temperature and its direct detection required cryogenically cooled devices, and also its efficient generation methods have been quite limited until recent years. The situation has changed drastically in the 1980s, when it was shown that the ultrafast femtosecond (fs) laser pulses can be used to relatively easily generate the THz waves as the form of monocycle pulses with broadband frequency spectrum, and to detect the temporal waveform of its electric field component directly in the time domain. This method, now referred to as THz time-domain spectroscopy (THz-TDS) [1, 2], is the basis for modern THz technologies, and has caused breakthrough in the spectroscopic study of this frequency region. Since its development, its usage has been spreading through both fundamental scientific study and industrial applications, and is considered one of the most important expanding areas of photonics researches. From the viewpoint of condensed-matter physics, lots of important elementary excitations are present in this frequency region, such as exciton, plasmon, phonon, molecular vibration, and magnon, etc. As such, the THz waves are suited to directly visualize the dynamics of resonant and impulsive responses of such

T. Kurihara, *Observation and Control of Magnetic Order Dynamics by Terahertz Magnetic Nearfield*, Springer Theses, https://doi.org/10.1007/978-981-16-8793-8_1

excitations. Owing to this feature, THz-TDS has become one of the most important tools in solid-state spectroscopy.

1.1.2 Usage of Intense THz Electric Fields

The light field can be used not only to probe the properties of solids, but also to cause drastic change of the system. Spectroscopy using intense electromagnetic waves provides rich information on the non-equillibrium states and dynamical behavior of matter. Owing to the recent developments in intense THz light sources [3–13], the nonlinear spectroscopy techniques in THz frequency region is rapidly growing. The electric fields reaching ~1 MV/cm that can be obtained in the state-of-the-art table-top intense THz-TDS systems is comparable to the strength of internal electric field that the electrons in solids feel due to the presence of ionic potentials. By using the laser-based THz sources, it is possible to apply such intense electric fields within picoseconds (ps). It should be noted that observation of the dynamical response of electronic systems under such high electric fields is quite difficult in conventional transport experiments where only the static electric fields are used, because under such strong electric fields, destructive changes of the system (i.e., electrical breakdown) occurs. Due to this unique feature of intense-THz spectroscopy, it has been applied to various studies such as ultrafast transport processes [14], excitonic transitions in solids [15–19], alignment of polar gas molecules [20–23], high-harmonics generation [24, 25], resonant excitation of Higgs amplitude mode in superconductors [26, 27] and even metal–insulator phase transition process in transition oxide materials [28].

1.1.3 Control of Spin Dynamics with THz Magnetic Fields

In addition to the above-mentioned successes in the usage of THz electric field, more recently, its *magnetic* counterpart is also attracting interest. Magnetic field component of light is usually regarded to cause negligible effects on the electronic systems in the visible region, because they are much too faster than the response time scale of the electron spins (magnetic permeability $\mu = 1$). However, the frequency of spin resonances in some magnetic systems such as antiferromagnets can reach as high as THz frequency and therefore, magnetic field of THz radiation can directly induce spin precession dynamics through magnetic dipole interaction ($\mu \neq 1$) [29–31]. Although many all-optical methods are known which enable the magnetization dynamics control, there exists clear advantage in the usage of THz magnetic fields. Due to its low photon energy, it enables purely-magnetic control of spin dynamics without causing excess heat and electronic transitions, which sometimes hinder the magnetic response.

Previously, it has been applied to probe the picosecond-fast spin precession dynamics in various magnetic materials. One of the most frequently used group of materials include the weak ferromagnet (for example, orthoferrites $RFeO_3$ [29, 32–37]), some ferromagnets (insulators such as ε-Fe_2O_3 [30] and metals [38, 39]) and anti-ferromagnet (NiO [31], MnO [40], etc.). As one of the earliest examples, in Ref. [30] it is demonstrated that the quasi single-cycle THz magnetic field can induce spin precession in picoseconds in a ferromagnet (ε-Fe_2O_3), resulting in the emission of circularly-polarized free-induction decay (FID) radiations. It was also shown that the use of two THz pump pulses can even coherently control the amplitude of spin precession motion, by either enhancing it or attenuating it depending on the arrival time of second pulse [29]. Similar demonstrations have been performed on various other materials as well [31, 35], suggesting the potential applicability of this method on ultrafast information processing devices in future.

In these previous studies, however, the amplitude of coherent spin precession motion induced by THz field was very small, reaching less than 1 degree in the best cases. Increase of the precession amplitude is expected to realize novel magnetization dynamics that are both important from the viewpoint of fundamental science and applications, namely, magnon-magnon scattering [41, 42], spin precession-induced second harmonics generation (SHG [43, 44]), parametric pumping [45–47] and ultrafast magnetization reversal [48–50], etc.

Motivated by this, experimental schemes to realize large-amplitude spin dynamics using intense THz magnetic fields have been proposed in recent theoretical studies. For example, in Ref. [49] it is theoretically suggested that the application of very intense THz magnetic fields on the antiferromagnetic spin systems may enable the permanent state of antiferromagnetic magnetization to be reversed on ps time scale. Some other works also report similar phenomenon, based on the theoretical calculation results. Though exciting these predictions are, realization of such direct magnetization reversal is challenging: the THz magnetic field strength required to realize such phenomena was estimated to be an order of ~10 T, which corresponds to the electric field strength of about ~30 MV/cm. This value is nearly 1–2 orders of magnitude larger than what is attainable by using the state-of-the-art table top intense THz pulse sources [12], and therefore, no one has realized such dynamics up to now.

1.1.4 Spin Reorientation Phase Transition (SRPT)

In order to break such situation and realize the drastic change in magnetic order using THz magnetic fields, in this Thesis we focus on a magnetic system which responds much sensitively to the magnetic field, namely, the phase transition process. Generally speaking, during the process of magnetic phase transition the spin systems undergo a state where its magnetic susceptibility diverges, i.e., softening behavior.

Previously, there has been some reports that the THz magnetic field has been applied to probe the static and dynamical behaviors of magnetic order during the phase transition process. As an example, Yamaguchi [33] demonstrated this in the orthoferrite $ErFeO_3$ which is a typical material known to exhibit spin reorientation phase transition (SRPT). SRPT is a special type of magnetic phase transition induced by the temperature-change of the magnetoanisotropy energy, wherein the antiferromagnetic order rotates by 90 degrees around the transition temperature region. It was indicated that by monitoring the FID emission in the THz-TDS measurement, it is possible to distinguish the low temperature (LT) and high temperature (HT) states of the system. Similar reports have been made on related materials with this method [32, 35]. However, the *control* of such magnetic transition process using THz fields has not been reported.

On the other hand, the control of SRPT process using all-optical methods go one step ahead compared to THz techniques. The circularly-polarized fs optical pulse can induce magnetic phase transition processes through ultrafast thermal [51–53] and nonthermal effects [54–63]. Furthermore, such ultrafast SRPT dynamics are shown to be sensitive to the small magnetic perturbations [64–67] applied at the initial state. For example, de Jong [66] suggested that even a small fluctuations of the magnetic system such as spin precession can lead to the formation of macroscopically aligned magnetization upon SRPT, breaking the symmetry of the phase transition path.

Considering such situation, it is expected that perturbation by THz field can cause a drastic change of the magnetic order if properly applied to the magnetic phase transition process assisted by photo irradiation. More precisely, the spin dynamics induced by the intense THz magnetic field may have a potential to determine the path of phase transition process initiated by the ultrafast optical heating. In Chaps. 4 and 5 such concept will be verified experimentally.

In order to realize the abovementioned magnetic field control of the phase transition process, we need to obtain the THz magnetic fields as strong as possible to apply on the spin system. For this purpose, we use the intense THz pulse with the peak electric field strength of several hundreds of kV/cm and magnetic field strength of several thousands of Oersted, produced from the state-of-the-art THz light source. We also introduce a method that is known to enable enhancement of electromagnetic field strengths by orders of magnitude: plasmonic enhancement in subwavelength metallic structures. Such techniques are conventionally used to enhance the energy of electric fields of light. In this Thesis, we apply it to enhance the *magnetic* field component of the THz waves, aiming at enhancing the amplitude of spin precession.

In addition to its ability of enhancing the absolute amplitude of the magnetic field strengths, plasmonics technique also has the advantage that in the nearfield regions, only the magnetic field can be selectively enhanced while keeping the electric field strength at small amplitude. This feature is preferable in the sense of realizing a purely *magnetic* control of the spin dynamics while avoiding the electric-field effects.

1.2 Field Enhancement Using Metallic Subwavelength Structures

1.2.1 Plasmonic Enhancement in Metallic Structures and Metamaterials

In this section, we briefly review the history and usage of plasmonic structures in spectroscopy and application. It has been already known in the 1950s that the electromagnetic energy of light can be confined in structured metallic surfaces, through collective excitation of electronic density wave coherently coupled to electric field of light (plasmons). This oscillation generates strongly localized and enhanced electric field in the vicinity (nearfield) of subwavelength metallic structures, typically reaching 10^2–10^3 of the incident light field. Such nearfield enhancement effects in structured metals have been applied from the 1980s for sensing technique such as, for example, the surface-enhanced Raman scattering measurement [68–71]. At those times it was discovered that the enhancement efficiency of electric field strength depends strongly on the subwavelength structures. Based on this large degree of freedom in the designability of such structures, the plasmonic enhancement effects are currently applied in wide area of photonic and optoelectronic devices such as chemical sensor [71], high-efficiency photovoltaic cells [72–76], fast-response Si photodiode [77, 78] and magnetooptical memory devices [53], etc.

Nowadays, the plasmonics study has evolved into a more generalized research concepts: *metamaterials*. The idea of metamaterials is to realize the electromagnetic response that does not exist in nature by artificially engineering the physical properties of materials—especially, their electric and magnetic responses to incident light field, i.e., dielectric constant ε and/or magnetic permeability μ of the material—using subwavelength structures. Inclusion of the concepts of artificial designing of ε and μ has extended the perspectives of photonics. Based on such concepts, many fascinating electromagnetic phenomena have been proposed and demonstrated so far, including negative refraction [79–83], super-resolution imaging [84], enhanced optical activity [85–87] and optical cloaking [88–91], etc.

1.2.2 Plasmonic Techniques in the THz Region

It has been considered until recently that such nearfield enhancements require the excitation of "true" plasmons in metals, whose frequencies are determined by intrinsic properties (carrier density) of host materials. However, it was shown in the 1990s that artificially designed subwavelength structures composed of ideal metals can also exhibit electromagnetic resonance modes which macroscopically shows a response similar to the true plasmons, at arbitrary frequencies depending on their structure designs. They are often called "spoof plasmons" [92–96]. Hence, the term *plasmonics* is now used to indicate the research field where both types of plasmons (the true- and spoof-plasmons) are used.

This extended the applicable frequency region of plasmonics and metamaterial techniques into the broad spectral region ranging from the VIS-UV down to microwave, millimeter and THz regions. Especially, their properties were extensively studied in the THz frequency region, because the wavelength (10 μm–10 mm) is much longer than the visible and thus sample fabrication is relatively easy. Also, in the THz region, conventional metals such as gold, copper, silver and aluminum that constitute metamaterials can be regarded as almost perfect conductor and therefore, it is suited for experimental implementation of the metamaterials with ideal characteristics.

1.2.3 Split-Ring Resonator as a Tool for THz Magnetic Field Enhancement

A lot of structure designs have been proposed to realize metamaterials in the THz region. One of the most frequently used structures is the split-ring resonator (SRR) [97–100]—a subwavelength metallic loop structure baring a gap. The feature of its lowest resonance mode is often modelled as inductance (L)—capacitance (C) circuit: the ring is regarded a single-turn coil L, and the gap is capacitance C. Because of the large magnetic moment induced by the current oscillation, the SRR can couple to the magnetic field of THz waves effectively and therefore has been widely used to realize artificial μ in the metamaterial studies.

In some cases, it is known that the magnetic field produced inside the ring associated with such resonance can reach several orders of magnitude larger than the original magnetic field contained in the incident light field. This allows us to use such magnetic nearfield to enhance the spin precession amplitude. One of the main concepts of the present Thesis is to realize such enhanced spin precession using SRR and to examine its dynamics, aiming at investigating the influence of coupling between the spin- and SRR-resonances.

1.3 Purpose of This Thesis and Outline

The ultimate purpose of this Thesis is to realize the macroscopic change of magnetization states using THz magnetic field. To do this, we propose the following two strategies:

(1) Realize the enhanced dynamics of spin precession using the magnetic field enhancement in SRR structure,

(2) Apply the enhanced magnetic field to a spin system which responds to external magnetic fluctuation highly sensitively (i.e., ultrafast magnetic phase transition process).

The following Chapters in this Thesis are presented as follows:

Chapter 2 introduces the properties of sample material used in the study (the erbium orthoferrite $ErFeO_3$), general spectroscopic techniques for generating/detecting the intense-THz pulses, designing of SRR structure using numerical simulation of electromagnetic fields.

Chapter 3 describes the result of experiment conducted to validate the concept of Strategy (1). It was shown that the SRR is capable of enhancing the amplitude of spin precession by resonant excitation through resonant coupling mediated by magnetic fields.

Chapter 4 is devoted to the result of experiment performed to realize Strategy (2), using the method demonstrated in Chapter 3. It was discovered for the first time that the THz magnetic field in SRR is capable of selectively controlling the path of phase transition, leading to the formation of macroscopically aligned ferromagnetic domain states.

Chapter 5 presents the detailed theoretical analysis for the results shown in Chap. 4.

Lastly, Chap. 6 summarizes the conclusion of this research.

References

1. Auston DH, Cheung KP, Smith PR (1984) Picosecond photoconducting Hertzian dipoles. Appl Phys Lett 45:284
2. Kida N, Murakami H, Tonouchi M (2005) In Sakaki K (ed) Terahertz optoelectronics. Springer-Verlag, Berlin
3. Hebling J, Yeh K-L, Hoffmann MC, Bartal B, Nelson KA (2008) Generation of high-power terahertz pulses by tilted-pulse-front excitation and their application possibilities. J Opt Soc Am B 25:B6–B19
4. Hirori H, Doi A, Blanchard F, Tanaka K (2011) Single-cycle terahertz pulses with amplitudes exceeding 1 MV/cm generated by optical rectification in $LiNbO_3$. Appl Phys Lett 98:091106
5. Fülöp JA, Pálfalvi L, Almási G, Hebling J (2010) Design of high-energy terahertz sources based on optical rectification. Opt Express 18:12311–12327
6. Hayashi S, Nawata K, Taira T, Shikata J-I, Kawase K, Minamide H (2014) Ultrabright continuously tunable terahertz-wave generation at room temperature. Sci Rep 4:5045
7. Vicario C, Monoszlai B, Hauri CP (2014) GV/m single-cycle terahertz fields from a laser-driven large-size partitioned organic crystal. Phys Rev Lett 112:213901
8. Shalaby M, Hauri CP (2015) Demonstration of a low-frequency three-dimensional terahertz bullet with extreme brightness. Nat Commun 6:5976
9. Bartel T, Gaal P, Reimann K, Woerner M, Elsaesser T (2005) Generation of single-cycle THz transients with high electric-field amplitudes. Opt Lett 30:2805
10. Minami Y, Nakajima M, Suemoto T (2011) Effect of preformed plasma on terahertz-wave emission from the plasma generated by two-color laser pulses. Phys Rev A 83:023828
11. Minami Y, Kurihara T, Yamaguchi K, Nakajima M, Suemoto T (2013) High-power THz wave generation in plasma induced by polarization adjusted two-color laser pulses. Appl Phys Lett 102:041105

12. Kampfrath T, Tanaka K, Nelson KA (2013) Resonant and nonresonant control over matter and light by intense terahertz transients. Nat Photon 7:680–690
13. Hwang HY et al (2015) A review of non-linear terahertz spectroscopy with ultrashort tabletop-laser pulses. J Mod Opt 62(18):1447–1479
14. Bowlan P, Kuehn W, Reimann K, Woerner M, Elsaesser T, Hey R, Flytzanis C (2011) High-field transport in an electron–hole plasma: transition from ballistic to drift motion. Phys Rev Lett 107:256602
15. Su FH, Blanchard F, Sharma G, Razzari L, Ayesheshim A, Cocker TL, Titova LV, Ozaki T, Kieffer J-C, Morandotti R, Reid M, Hegmann FA (2009) Terahertz pulse induced intervalley scattering in photoexcited GaAs. Opt Express 17:9620–9629
16. Ho I-C, Zhang X-C (2011) Driving intervalley scattering and impact ionization in InAs with intense terahertz pulses. Appl Phys Lett 98:241908
17. Wen H, Wiczer M, Lindenberg AM (2008) Ultrafast electron cascades in semiconductors driven by intense femtosecond terahertz pulses. Phys Rev B 78:125203
18. Hoffmann MC, Hebling J, Hwang HY, Yeh K-L, Nelson KA (2009) Impact ionization in InSb probed by terahertz pump–terahertz probe spectroscopy. Phys Rev B 79:161201(R)
19. Hirori H, Shinokita K, Shirai M, Tani S, Kadoya Y, Tanaka K (2011) Extraordinary carrier multiplication gated by a picosecond electric field pulse. Nat Commun 2:594
20. Fleischer S, Zhou Y, Field RW, Nelson KA (2011) Molecular orientation and alignment by intense single-cycle THz pulses. Phys Rev Lett 107:163603
21. Fleischer S, Field RW, Nelson KA (2012) Commensurate two-quantum coherences induced by time-delayed THz fields. Phys Rev Lett 109:123603
22. Liu J, Zhang X-C (2009) Terahertz-radiation-enhanced emission of fluorescence from gas plasma. Phys Rev Lett 103:235002
23. Shalaby M, Hauri CP (2015) Air nonlinear dynamics initiated by ultra-intense lambda-cubic terahertz pulses. Appl Phys Lett 106:181108
24. Schubert O, Hohenleutner M, Langer F, Urbanek B, Lange C, Huttner U, Golde D, Meier T, Kira M, Koch SW, Huber R (2014) Sub-cycle control of terahertz high-harmonic generation by dynamical Bloch oscillations. Nat Photon 8:119
25. Higuchi T, Stockman MI, Hommelhoff P (2014) Strong-field perspective on high-harmonic radiation from bulk solids. Phys Rev Lett 113:213901
26. Matsunaga R, Hamada YI, Makise K, Uzawa Y, Terai H, Wang Z, Shimano R (2013) Higgs amplitude mode in the BCS superconductors Nb1-xTixN induced by terahertz pulse excitation. Phys Rev Lett 111:057002
27. Matsunaga R, Tsuji N, Fujita H, Sugioka A, Makise K, Uzawa Y, Terai H, Wang Z, Aoki H, Shimano R (2014) Light-induced collective pseudospin precession resonating with Higgs mode in a superconductor. Science 345(6201):1145–1149
28. Liu M et al (2012) Terahertz-field-induced insulator-to-metal transition in vanadium dioxide metamaterial. Nature 487:345–348
29. Yamaguchi K, Nakajima M, Suemoto T (2010) Coherent control of spin precession motion with impulsive magnetic fields of half-cycle terahertz radiation. Phys Rev Lett 105:237201
30. Nakajima M, Namai A, Ohkoshi S, Suemoto T (2010) Ultrafast time domain demonstration of bulk magnetization precession at zero magnetic field ferromagnetic resonance induced by terahertz magnetic field. Opt Exp 18(17):18260
31. Kampfrath T, Sell A, Klatt G, Pashkin A, Mährlein S, Dekorsy T, Wolf M, Fiebig M, Leitenstorfer A, Huber R (2011) Coherent terahertz control of antiferromagnetic spin waves. Nat Photon 5:31–34
32. Suemoto T, Nakamura K, Kurihara T, Watanabe H (2015) Magnetization-free measurements of spin orientations in orthoferrites using terahertz time domain spectroscopy. Appl Phys Lett 107:042404
33. Yamaguchi K, Kurihara T, Minami Y, Nakajima M, Suemoto T (2013) Terahertz time-domain observation of spin reorientation in orthoferrite $ErFeO_3$ through magnetic free induction decay. Phys Rev Lett 110:137204

34. Reid AHM, Rasing Th, Pisarev RV, Dürr HA, Hoffmann MC (2015) Terahertz-driven magnetism dynamics in the orthoferrite $DyFeO_3$. Appl Phys Lett 106:082403
35. Song G, Jiang J, Kang B, Zhang J, Cheng Z, Ma G, Cao S (2015) Spin reorientation transition process in single crystal $NdFeO_3$. Solid Stat Commun 211:47–51
36. Fu X, Liu X, Zhou J (2014) Terahertz spectroscopic observation of spin reorientation induced antiferromagnetic mode softening in $DyFeO_3$ ceramics. Mater Lett 132:190–192
37. Mikhaylovskiy RV, Hendry E, Secchi A, Mentink JH, Eckstein M, Wu A, Pisarev RV, Kruglyak VV, Katsnelson MI, Rasing Th, Kimel AV (2015) Ultrafast optical modification of exchange interactions in iron oxides. Nat Commun 6:8190
38. Wang Z, Pietz M, Walowski J, Förster A, Lepsa MI, Münzenberg M (2008) Spin dynamics triggered by subterahertz magnetic field pulses. J Appl Phys 103:123905
39. Vicario C, Ruchert C, Ardana-Lamas F, Derlet PM, Tudu B, Luning J, Hauri CP (2013) Off-resonant magnetization dynamics phase-locked to an intense phase-stable terahertz transient. Nat Photon 7
40. Moriyasu T, Wakabayashi S, Kohmoto T (2013) Observation of antiferromagnetic magnons and magnetostriction in manganese oxide using terahertz time-domain spectroscopy. J Infrared Milli Terahz Waves 34:277–288
41. Schultheiss H, Vogt K, Hillebrands B (2013) Direct observation of nonlinear four-magnon scattering in spin-wave microconduits. Phys Rev B 86:054414
42. L'vov VS, Starobinets SS (1971) Nonlinear theory of ferromagnetic resonance. Sov Phys Solid Stat 13:2
43. Boscaino R, Ciccarello I, Cusumano C, Strandberg MWP (1971) Second-harmonic generation and spin decoupling in resonant two-level spin systems. Phys Rev B 3:8
44. Boscaino R, Ciccarello I, Cusumano C (1968) Second-harmonic generation by a resonant spin system. Phys Rev Lett 20:9
45. Gerrits T, Krivosik P, Schneider ML, Patton CE, Silva TJ (2007) Direct detection of nonlinear ferromagnetic resonance in thin films by the magneto-optical Kerr effect. Phys Rev Lett 98:207602
46. Suhl H (1957) The theory of ferromagnetic resonance at high signal powers. Phys Chem Solids 1:209–227
47. Le Gall H (1968) Quantum-mechanical treatment of the transient processes with three-magnon non-linear transitions under perpendicular pumping. Phys. Stat. Sol. 28:495
48. Shalaby M, Vidal F, Peccianti M, Morandotti R, Enderli F, Feurer T, Patterson BD (2013) Terahertz macrospin dynamics in insulating ferrimagnets. Phys Rev B 88:140301(R)
49. Wienholdt S, Hinzke D, Nowak U (2012) THz switching of antiferromagnets and ferrimagnets. Phys Rev Lett 108:247207
50. Mochizuki M, Nagaosa N (2010) Theoretically predicted picosecond optical switching of spin chirality in multiferroics. Phys Rev Lett 105:147202
51. Stanciu CD, Tsukamoto A, Kimel AV, Hansteen F, Kirilyuk A, Itoh A, Rasing Th (2007) Subpicosecond magnetization reversal across ferrimagnetic compensation points. Phys Rev Lett 99:217204
52. Challener WA, Peng C, Itagi AV, Karns D, Peng W, Peng Y, Yang X, Zhu X, Gokemeijer NJ, Hsia Y-T, Ju G, Rottmayer RE, Seigler MA, Gage EC (2009) Heat-assisted magnetic recording by a near-field transducer with efficient optical energy transfer. Nat Photon 3:220–224
53. Zhou N, Xu X, Hammack AT, Stipe BC, Gao K, Scholz W, Gage EC (2014) Plasmonic near-field transducer for heat-assisted magnetic recording. Nanophotonics 3(3):141–155
54. Stanciu CD, Hansteen F, Kimel AV, Kirilyuk A, Tsukamoto A, Itoh A, Rasing Th (2007) All-optical magnetic recording with circularly polarized light. Phys Rev Lett 99:047601
55. Kimel AV, Kirilyuk A, Usachev PA, Pisarev RV, Balbashov AM, Rasing Th (2005) Ultrafast non-thermal control of magnetization by instantaneous photomagnetic pulses. Nature 435:655–657
56. Nishitani J, Kozuki K, Nagashima T, Hangyo M (2010) Terahertz radiation from coherent antiferromagnetic magnons excited by femtosecond laser pulses. Appl Phys Lett 96:221906

57. Kimel AV, Kirilyuk A, Rasing T (2007) Femtosecond opto-magnetism: ultrafast laser manipulation of magnetic materials. Laser Photon Rev 1(3):275–287
58. Kurkin MI, Bakulina NB, Pisarev RV (2008) Transient inverse Faraday effect and ultrafast optical switching of magnetization. Phys Rev B 78:134430
59. Gridnev VN (2008) Phenomenological theory for coherent magnon generation through impulsive stimulated Raman scattering. Phys Rev B 77:094426
60. Kirilyuk A, Kimel AV, Rasing T (2013) Laser-induced magnetization dynamics and reversal in ferrimagnetic alloys. Rep Prog Phys 76:026501
61. Vahaplar K, Kalashnikova AM, Kimel AV, Hinzke D, Nowak U, Chantrell R, Tsukamoto A, Itoh A, Kirilyuk A, Rasing T (2009) Ultrafast path for optical magnetization reversal via a strongly nonequilibrium state. Phys Rev Lett 103:117201
62. Satoh T, Duong NP, Fiebig M (2006) Coherent control of antiferromagnetism in NiO. Phys Rev B 74:012404
63. Akimoto R, Ando K, Sasaki F, Kobayashi S, Tani T (1998) Optical control of Larmor precession of Mn^{2+} moments in $CdTe/Cd_{1-x}Mn_xTe$ quantum wells. J Appl Phys 84:6318
64. Kimel AV, Kirilyuk A, Tsvetkov A, Pisarev RV Rasing Th (2004) Laser-induced ultrafast spin reorientation in the antiferromagnet $TmFeO_3$. Nature 429:850-853
65. Kimel AV, Stanciu CD, Usachev PA, Pisarev RV, Gridnev VN, Kirilyuk A, Rasing Th Optical excitation of antiferromagnetic resonance in $TmFeO_3$. Phys. Rev. B (R) 74:060403
66. de Jong JA, Razdolski I, Kalashnikova AM, Pisarev RV, Balbashov AM, Kirilyuk A, Rasing Th, Kimel AV (2012) Coherent control of the route of an ultrafast magnetic phase transition via low-amplitude spin precession. Phys Rev Lett 108:157601
67. de Jong JA, Kimel AV, Pisarev RV, Kirilyuk A, Rasing Th (2011) Laser-induced ultrafast spin dynamics in $ErFeO_3$. Phys Rev B 84:104421
68. Anker JN, Paige Hall W, Lyandres O, Shah NC, Zhao J, Van Duyne RP (2008) Biosensing with plasmonic nanosensors. Nat Mater 7:442–453
69. Mulvaney P (1996) Surface plasmon spectroscopy of nanosized metal particles. Langmuir 12(3):788–800
70. Barnes WL, Dereux A, Ebbesen TW (2003) Surface plasmon subwavelength optics. Nature 424:824–830
71. Willets KA, Van Duyne RP (2007) Localized surface plasmon resonance spectroscopy and sensing. Annu Rev Phys Chem 58:267–297
72. Rand BP, Peumans P, Forrest SR (2004) Long-range absorption enhancement in organic tandem thin-film solar cells containing silver nanoclusters. J Appl Phys 96:7519
73. Morfa AJ, Rowlen KL, Reilly TH III, Romero MJ, van de Lagemaat J (2008) Plasmon-enhanced solar energy conversion in organic bulk heterojunction photovoltaics. Appl Phys Lett 92:013504
74. Beck FJ, Polman A, Catchpole KR (2009) Tunable light trapping for solar cells using localized surface plasmons. J Appl Phys 105:114310
75. Pillain S, Green MA (2010) Plasmonics for photovoltaic applications. Sol Energy Mater Sol Cells 94:1481–1486
76. Clavero C (2014) Plasmon-induced hot-electron generation at nanoparticle/metal-oxide interfaces for photovoltaic and photocatalytic devices. Nat Photon 8:95–103
77. Ishi T, Fujikata J, Makita K, Baba T, Ohashi K (2005) Si nano-photodiode with a surface plasmon antenna. Jpn J Appl Phys 44(2):12–15
78. Le Perchec J, Desieres Y, Espiau de Lamaestre R (2009) Plasmon-based photosensors comprising a very thin semiconducting region. Appl Phys Lett 94:181104
79. Valentine J, Zhang S, Zentgraf T, Ulin-Avila E, Genov DA, Bartal G, Zhang X (2008) Three-dimensional optical metamaterial with a negative refractive index. Nature 455:376–379
80. Shalaev VM, Cai W, Chettiar UK, Yuan H-K, Sarychev AK, Drachev VP, Kildishev AV (2005) Negative index of refraction in optical metamaterials. Opt Lett 30(24):3356–3358
81. Shelby RA, Smith DR, Schultz S (2001) Experimental verification of a negative index of refraction. Science 292

82. Pendry JB, Schurig D, Smith DR (2006) Controlling electromagnetic fields. Science 312(5781):1780–1782
83. Padilla WJ, Basov DN, Smith DR (2006) Negative refractive index metamaterials. Mater Today 9(7–8)
84. Pendry JB (2000) Negative refraction makes a perfect lens. Phys Rev Lett 85:3966
85. Kuwata-Gonokami M, Saito N, Ino Y, Kauranen M, Jefimovs K, Vallius T, Turunen J, Svirko Y (2005) Giant optical activity in quasi-two-dimensional planar nanostructures. Phys Rev Lett 95:227401
86. Kanda N, Konishi K, Kuwata-Gonokami M (2009) Light-induced terahertz optical activity. Opt Lett 34(19)
87. Zhou J, Chowdhury DR, Zhao R, Azad AK, Chen H-T, Soukoulis CM, Taylor AJ, O'Hara JF (2012) Terahertz chiral metamaterials with giant and dynamically tunable optical activity. Phys Rev B 86:035448
88. Schurig D, Mock JJ, Justice BJ, Cummer SA, Pendry JB, Starr AF, Smith DR (2006) Metamaterial electromagnetic cloak at microwave frequencies. Science 314(5801):977–980
89. Chen H, Wu B-I, Zhang B, Kong JA (2007) Electromagnetic wave interactions with a metamaterial cloak. Phys Rev Lett 99:063903
90. Cai W, Chettiar UK, Kildishev AV, Shalaev VM (2007) Optical cloaking with metamaterials. Nat Photon 1:224–227
91. Alù A, Engheta N (2008) Multifrequency optical invisibility cloak with layered plasmonic shells. Phys Rev Lett 100:113901
92. Pendry JB, Martín-Moreno L, García-Vidal FJ (2004) Mimicking surface plasmons with structured surfaces. Science 305:847–848
93. García-Vidal FJ, Martín-Moreno L, Pendry JB (2005) Surfaces with holes in them: New plasmonic metamaterials. J Opt A: Pure Appl Opt 7:S97–S101
94. Maier SA, Andrews SR, Martín-Moreno L, García-Vidal FJ (2006) Terahertz surface plasmon-polariton propagation and focusing on periodically corrugated metal wires. Phys Rev Lett 97:176805
95. Shen L, Chen X, Yang T-J (2008) Terahertz surface plasmon polaritons on periodically corrugated metal surfaces. Opt Exp 16(5):3326–3333
96. Yu N, Wang QJ, Kats MA, Fan JA, Khanna SP, Li L, Giles Davies A, Linfield EH, Capasso F (2010) Designer spoof surface plasmon structures collimate terahertz laser beams. Nat Mater 9:730–735
97. Yen TJ, Padilla WJ, Fang N, Vier DC, Smith DR, Pendry JB, Basov DN, Zhang X (2004) Terahertz magnetic response from artificial materials. Science 303:1494
98. Katsarakis N, Koschny T, Kafesaki M, Economou EN, Soukoulis CM (2004) Electric coupling to the magnetic resonance of split ring resonators. Appl Phys Lett 84:15
99. Padilla WJ, Taylor AJ, Highstrete C, Lee M, Averitt RD (2006) Dynamical electric and magnetic metamaterial response at terahertz frequencies. Phys Rev Lett 96:107401
100. Anantha Ramakrishna S (2005) Physics of negative refractive index materials. Rep Prog Phys 68:449–521

Chapter 2
Background

Abstract In this chapter, theoretical backgrounds relevant for the experiments performed in the subsequent chapters are presented in detail. Firstly, in Sect. 2.1, basic physical properties of the sample material used in the study (rare-earth orthoferrite $ErFeO_3$) are reviewed, accompanied by theoretical model used to describe its magnetic properties. Secondly, in Sect. 2.2 the principles of experimental setups used to generate and detect the intense THz pulses are presented. Additionally, magnetooptical effects used for the detection of spin dynamics are presented. Lastly, in Sect. 2.3 the basic features of split-ring resonator (SRR) metamaterials and its field enhancement effects are reviewed. Also, we refer to previous researches focused on the coupling between SRR and spin systems.

Keywords Orthoferrite · Spin dynamics · Intense terahertz pulse · Magnetooptical effect · Split ring resonator

2.1 Rare-earth Orthoferrites ($RFeO_3$)

2.1.1 Basic Properties of Orthoferrite

Sample material used in the study is erbium orthoferrite ($ErFeO_3$), which belongs to the series of rare-earth orthoferrite $RFeO_3$, where R = rare earth ions [1]. From chemical composition they can be categorized as perovskite iron oxides. As its name indicates, the crystal structure of $RFeO_3$ (Fig. 2.1) is slightly distorted from cubic and exhibits orthorhombic symmetry (space group = Pbnm). The crystal field splitting of the 3d orbits of the Fe^{3+} ion (i.e., 10 Dq) is not large, and so the five electrons in Fe^{3+} ion are packed in accordance with Hund's rule and exhibit high-spin state ($S = 5/2$) as the ground state. Like many other iron-oxide magnetic materials (ferrites), the 3d electron orbit of the Fe^{3+} ion contained in $RFeO_3$ has an overlap with the electron orbit of the legand ions (3p electron orbit of O^{2-} ions that surrounds Fe^{3+} ion octahedrally) and interacts with the neighboring Fe^{3+} ions via superexchange. This causes antiferromagnetic ordering of the Fe^{3+} system [2] below Néel temperature T_N ~600 K. Furthermore, the orthorhombic distortion of the crystal symmetry induces nonequivalence of the Fe^{3+} sites. This leads to the antisymmetric exchange

T. Kurihara, *Observation and Control of Magnetic Order Dynamics by Terahertz Magnetic Nearfield*, Springer Theses, https://doi.org/10.1007/978-981-16-8793-8_2

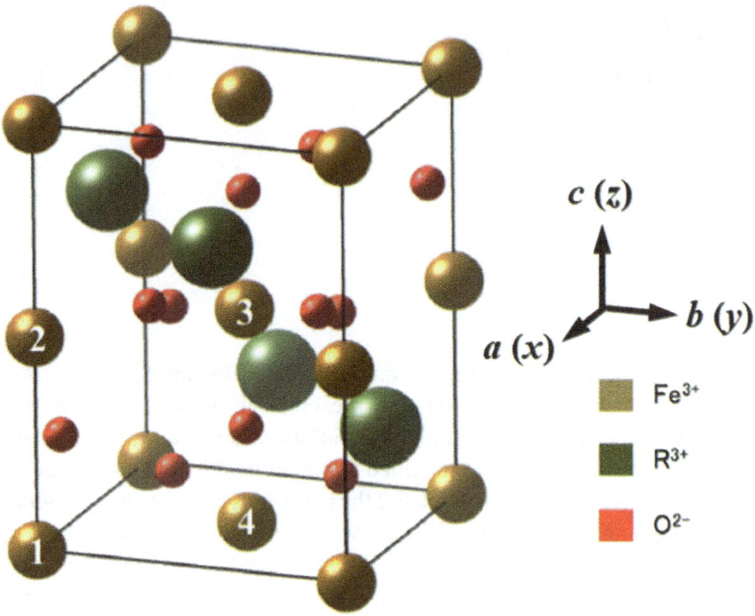

Fig. 2.1 Crystal structure of the rare-earth orthoferrite $RFeO_3$

interaction called Dzyaloshinskii-Moriya (DM) interaction [3, 4], and due to this, the antiferromagnetically aligned spins are slightly canted from each other, giving rise to the appearance of weak ferromagnetic moments.

2.1.2 Spin Configuration

Here, we will review the spin configuration of $RFeO_3$ in the equilibrium state. It is known that in a unit cell, four nonequivalent Fe^{3+} sites are present. Figure 2.2 schematically illustrates the three antiferromagnetic spin configurations that are allowed under the crystal symmetry of Pbnm [5]. Using the notation introduced by Bertaut [2], they are commonly called Γ_1, Γ_2 and Γ_4 phases. While there are actually four spins, in most cases it is approximated that $s_1 = s_3$ and $s_2 = s_4$. This simplification is called "two-spin approximation". It is known that such simplification has negligible effects in the consideration of basic feature of the magnetic resonance and ground states of orthoferrites, and are frequently used to describe their properties [5, 6]. Therefore, we will continue our argument under this approximation hereafter. Then, using the ferromagnetic vector $\boldsymbol{M} = \boldsymbol{s}_1 + \boldsymbol{s}_2$ and antiferromagnetic vector $\boldsymbol{G} = \boldsymbol{s}_1 - \boldsymbol{s}_2$, the abovementioned spin configurations can be characterized by the directions of $\boldsymbol{M}$ and $\boldsymbol{G}$. For example, $\Gamma_4 = \Gamma_4\ (G_x M_z)$, which means that the spins are

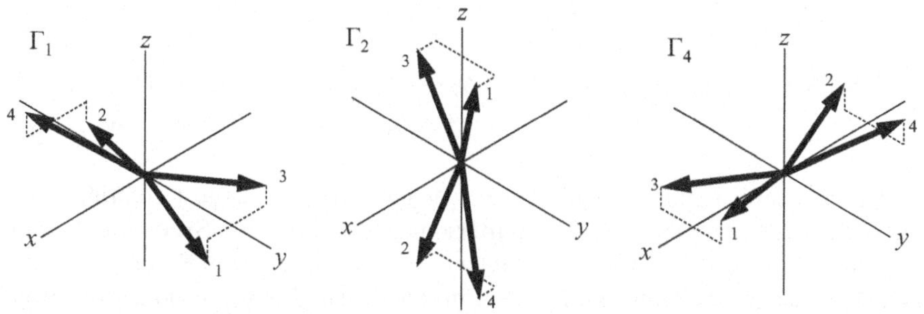

Fig. 2.2 Magnetic configurations in orthoferrites allowed under crystal symmetry and antiferromagnetic interaction

aligned along x-axis and the macroscopic magnetization is directed towards z-axis. Likewise, notation $\Gamma_2 = \Gamma_2\ (G_y M_x)$ and $\Gamma_1 = \Gamma_1\ (G_y)$ are often used.

In many orthoferrites where the rare-earth ions are magnetic, temperature-induced phase transition between Γ_1 and Γ_4, or Γ_2 and Γ_4 are known to occur. These phase transitions are called spin reorientation phase transition (SRPT). The transition between Γ_2 and Γ_4 is called "rotation-type" SRPT, because the spin configuration continuously rotates by 90° from Γ_2 (in the low temperature (LT)) to Γ_4 (in the high temperature (HT)) within some finite temperature regions. In between LT and HT, spin configuration exhibits an intermediate state Γ_{24}, wherein the spins are rotated by 0–90° depending on temperature. This type of SRPT occurs in many orthoferrites, such as $NdFeO_3$, $SmFeO_3$, $TmFeO_3$, $HoFeO_3$, $YbFeO_3$ and also in $ErFeO_3$ [7–9], which is the sample material used in this Thesis. On the other hand, the transition between Γ_1 and Γ_4 is quite rare in orthoferrite and occurs only in $DyFeO_3$. Unlike the previous case, this transition occurs abruptly at some temperature T_r, and for that it is called "abrupt-type" SRPT, or Morin transition [10].

It is recognized that the origin of the abovementioned SRPT phenomena is ascribed to the change of Fe^{3+} ion's magnetic anisotropy energy against temperature, owing to the temperature change of rare-earth ion's paramagnetic magnetization that is coupled to the Fe^{3+} ion through exchange interaction [11, 12]. In the next subsection, we present the free energy description of the spin system that is most frequently used to express the SRPT phenomena in $RFeO_3$.

2.1.3 Free-Energy Description of the Rotation-Type SRPT

The rotation-type SRPT process that takes place in the $ErFeO_3$ is known to be well described by the free energy of the abovementioned two-spin model in the following manner:

$$F(\boldsymbol{S}_1, \boldsymbol{S}_2) = J(\boldsymbol{S}_1 \cdot \boldsymbol{S}_2) + \boldsymbol{D} \cdot (\boldsymbol{S}_1 \times \boldsymbol{S}_2) + A_{xx}\left(S_{1x}^2 + S_{2x}^2\right) + A_{zz}\left(S_{1z}^2 + S_{2z}^2\right) \\ + A_4\left(S_{1x}^4 + S_{1y}^4 + S_{1z}^4 + S_{2x}^4 + S_{2y}^4 + S_{2z}^4\right) + \boldsymbol{H}_0 \cdot (\boldsymbol{S}_1 + \boldsymbol{S}_2) \quad (2.1)$$

The first term is the exchange energy, which gives the largest contribution to the free energy and constructs the antiferromagnetic structure. Second term is the Dzyaloshinskii-Moriya interaction that tends to make the angle of S_1 and S_2 perpendicular to each other. Third and fourth terms are the single-ion anisotropy of second and fourth order. The last term is the magnetic dipole energy caused by the external magnetic field (i.e., static or THz magnetic field). The magnitudes of each parameter are deduced from various experiments, and estimated to be J ~600 K, D/J–O (10^{-2}), $A_{xx,zz}$/J–O (10^{-3}) and A_4/J–O (10^{-4}) [13, 14].

The cant angle β caused by Dzyaloshinskii-Moriya interaction is known to be almost temperature independent, and is given by:

$$\beta \sim (1/2)\tan^{-1}(D/J) \sim O(1\,\text{degree}) \quad (2.2)$$

During the SRPT, the equilibrium position of $\mathbf{S}_1$ and $\mathbf{S}_2$ spins lies within in the x–z plane, as shown in Fig. 2.3a. We can then describe this rotation by substituting;

$$S_1 = (\sin(\theta + \beta), 0, \cos(\theta + \beta)) \text{ and } S_2 = (-\sin(\theta - \beta), 0, -\cos(\theta - \beta)), \quad (2.3)$$

and rewrite (Eq. 2.1) as a function of θ as:

$$F(\theta) = K_u \cos(2\theta) + K_b \cos(4\theta), \quad (2.4)$$

where

$$K_u = -4(A_{xx} - A_{zz})\cos 2\beta,\ K_b = A_4 \cos 4\beta. \quad (2.5)$$

It is known that the temperature dependence of AFM- and FM-mode resonance frequencies near the SRPT temperature regions can be approximated by assuming A_{xx} (T) = constant, A_4 (T) = constant and A_{zz} (T) = linear against temperature. Therefore, we can substitute them into (Eq. 2.5) to obtain

$$K_u(T) = K_0 + K_1 T \quad (2.6)$$

By substituting this into (Eq. 2.4), the free energy curves of Fe^{3+} spin system at various temperatures can be obtained, as shown in Fig. 2.3b. As can be seen, the potential minimum is located at $\theta = 0$ (**M** // a) in the low temperature (LT) phase, $\theta = \pm 90$ (**M** // $\pm$ c) in the high temperature (HT) phase, and at $0 < |\theta| < 90$ (two minimum points around $\theta = 0$) in the intermediate temperature region. From this,

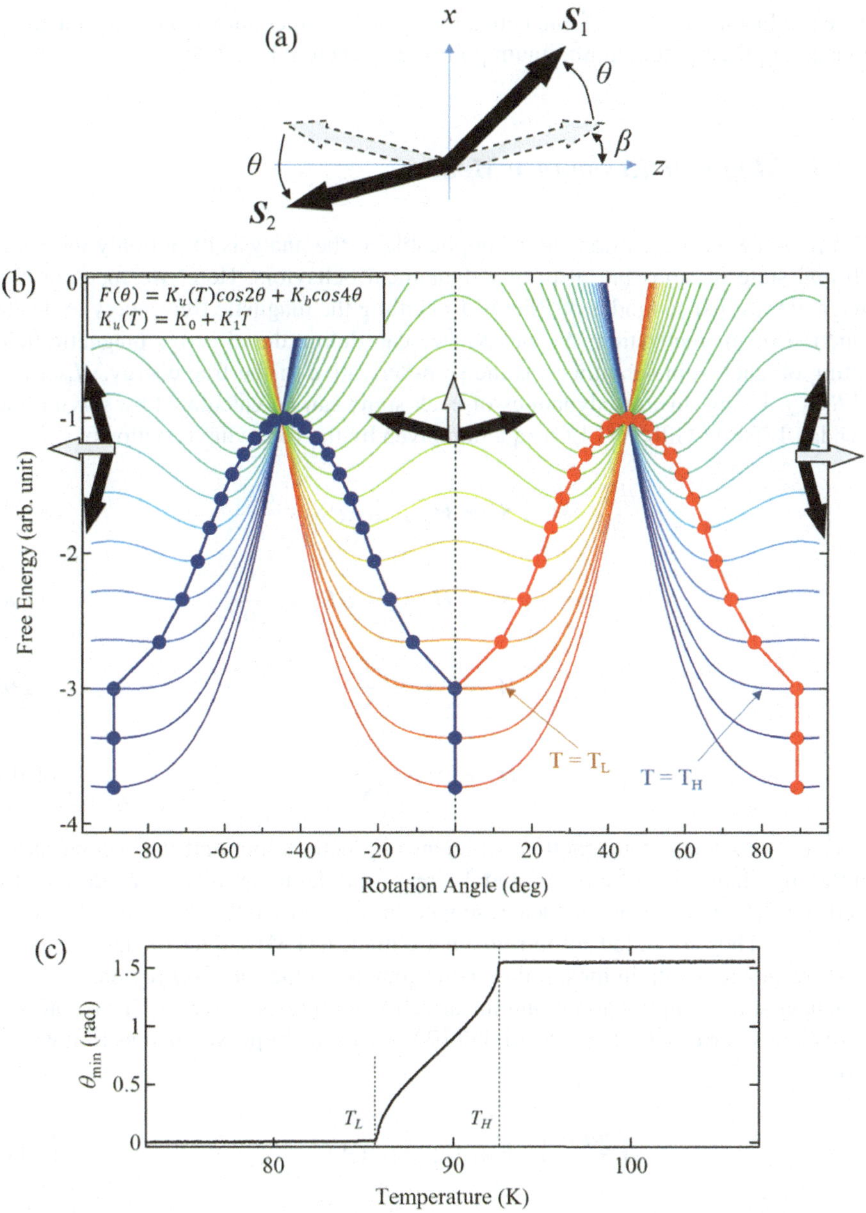

Fig. 2.3 **a** Definition of θ and β. **b** Free energy calculated from Eq. 1.5. Filled circles indicate the points where F (θ) takes its minimum. **c** The free energy minimum angle θ_{min} as a function of temperature

the equilibrium direction of magnetization at each temperature can be calculated by connecting their potential minimum points, as shown in Fig. 2.3c.

2.1.4 Magnetic Resonance Modes

The free-energy description can be applicable to the analysis of not only the equilibrium state of spins but also their dynamical behaviors. Here, we focus on the magnetic resonance modes of $RFeO_3$. Regarding the magnetic free energy to be the function of magnetizations $\boldsymbol{S}_1$ and $\boldsymbol{S}_2$, we can define the effective magnetic field acting on each spin sublattices as the spin-derivative of the free energy: $\boldsymbol{H}_{\mathrm{eff}\,1,2} = \partial F/\partial \boldsymbol{S}_{1,2}$. Using this, the dynamics of each spins can be calculated by solving the Landau-Lifshitz-Gilbert (LLG) equation given by the following equations:

$$\frac{d\boldsymbol{S}_1}{dt} = -\gamma \boldsymbol{S}_1 \times \boldsymbol{H}_{eff,1} + \alpha \boldsymbol{S}_1 \times \frac{d\boldsymbol{S}_1}{dt} \tag{2.7}$$

$$\frac{d\boldsymbol{S}_2}{dt} = -\gamma \boldsymbol{S}_2 \times \boldsymbol{H}_{eff,2} + \alpha \boldsymbol{S}_2 \times \frac{d\boldsymbol{S}_2}{dt} \tag{2.8}$$

$$\boldsymbol{H}_{eff,1} = -\frac{\partial F}{\partial \boldsymbol{S}_1} \tag{2.9}$$

$$\boldsymbol{H}_{eff,2} = -\frac{\partial F}{\partial \boldsymbol{S}_2} \tag{2.10}$$

Here, γ is the gyromagnetic coefficient of electron spin and the second terms in the right hand side of eqs. 2.7 and 2.8 represent damping. It was first derived by Herrmann [6] that two independent resonance modes exist as the solution to this set of equations. They are called antiferromagnetic-mode (AFM) and ferromagnetic-mode (FM) resonances, and in most orthoferrites they lie in the sub-THz region.

Under some simplifications and linearization with respect to $\mathbf{S}_{1,2}$, The resonance frequencies corresponding to AFM and FM modes are approximated as follows [6, 15]:

$$\left(\frac{\omega_{FM}}{\gamma}\right)^2 = 4(J + A_{xx})(A_{xx} - A_{zz}) \tag{2.11}$$

$$\left(\frac{\omega_{AFM}}{\gamma}\right)^2 = 4JA_{xx} + 4A_{xx}(A_{xx} - A_{zz}) + D^2 \tag{2.12}$$

The motions of $\boldsymbol{S}_1$, $\boldsymbol{S}_2$ and total magnetic moment $\boldsymbol{M} = \boldsymbol{S}_1 + \boldsymbol{S}_2$ are schematically shown in Fig. 2.4 [6]. The FM involves the precession motion of macroscopic magnetization $\boldsymbol{M}$, for which it is called "ferromagnetic" mode. On the other hand, in the AFM the stretching motion of $\boldsymbol{M}$ occurs instead of precession. These two

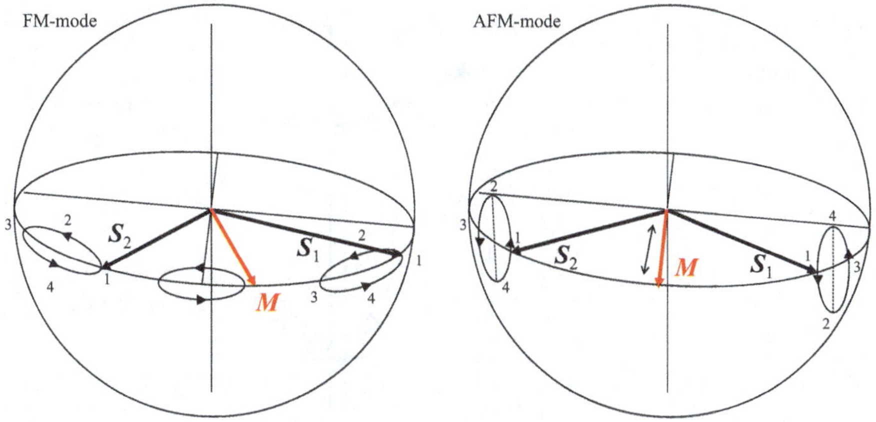

Fig. 2.4 The two magnetic resonance modes of orthoferrites in the THz region. The ferromagnetic (FM)-mode corresponds to the precession motion of $\boldsymbol{M}$, while antiferromagnetic (AFM)-mode can be regarded as its stretching motion

modes are optically active via magnetic dipole interaction, and can be excited by the application of THz magnetic fields with properly aligned polarization. In the Γ_4 state, for example, the AFM is excited when H_{THz} // z (magnetic field is parallel to $\boldsymbol{M}$) and FM is excited when $\boldsymbol{H}_{THz}$ // x and y (magnetic field is perpendicular to $\boldsymbol{M}$). In this Thesis we focus on the dynamics of FM-mode resonance.

2.1.5 Temperature Dependence of FM Mode Resonance

As mentioned in the previous subsection, the frequencies of AFM and FM mode resonance are dependent on the magnetoanisotropy coefficients $A_{zz,xx}$. Especially, that of FM mode is proportional to $(A_{zz} - A_{xx})^{1/2}$ [15]. In $ErFeO_3$, A_{zz} and A_{xx} are almost matched with each other near the SRPT temperature regions, and therefore, the FM frequency softens around this temperature. The temperature dependence of FM-mode mode frequency is shown in Fig. 2.5 [16].

2.2 Generation and Detection of THz Pulses

In this section the general principles of generation and detection methods of THz pulses using nonlinear optical processes are introduced.

Fig. 2.5 The temperature-dependence of FM- and AFM modes of $ErFeO_3$ [16]

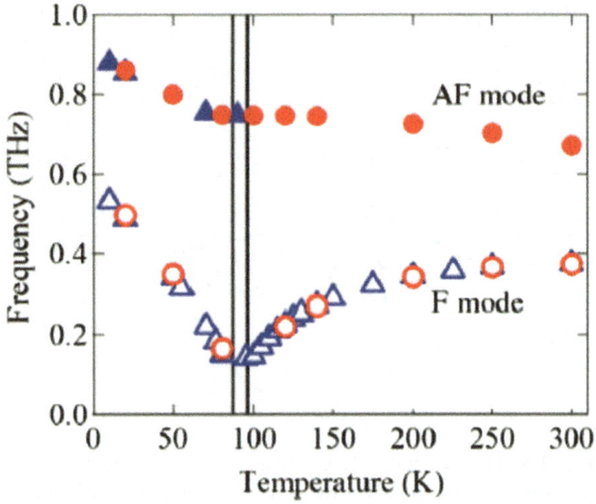

2.2.1 THz Time-Domain Spectroscopy (THz-TDS)

The uniqueness of THz-TDS is that the waveform of electric field is directly measured in the time-domain. This means that not only the magnitude but also the real and imaginary parts of optical properties of matter such as complex refractive index and absorbance can be directly measured without using Kramers–Kronig relations, unlike conventional spectroscopic techniques in the visible regions which measure the intensity of light. When applied to the magnetic materials, it enables the measurement of spin dynamics in the time domain.

General concept of THz-TDS is shown schematically in Fig. 2.6. In THz-TDS experiments, femtosecond laser pulses (typical time width = 10–100 fs) are used for both the generation and detection of THz waves. Various materials are known

Fig. 2.6 General concept of THz-TDS. The electric field causes a change of dielectric properties proportional to its field amplitude in a detector material, which is probed by a subsequently incident femtosecond laser pulse

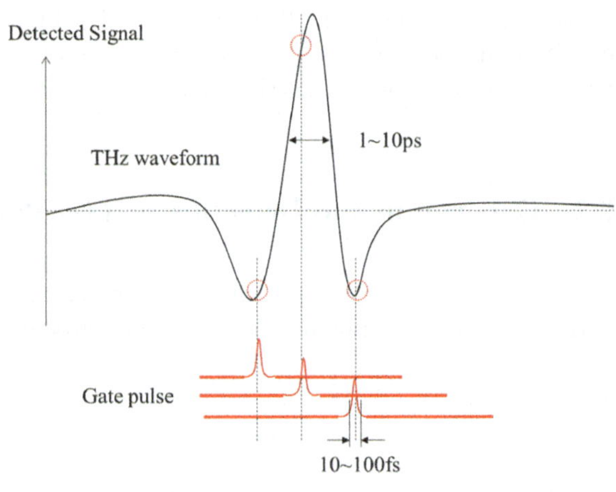

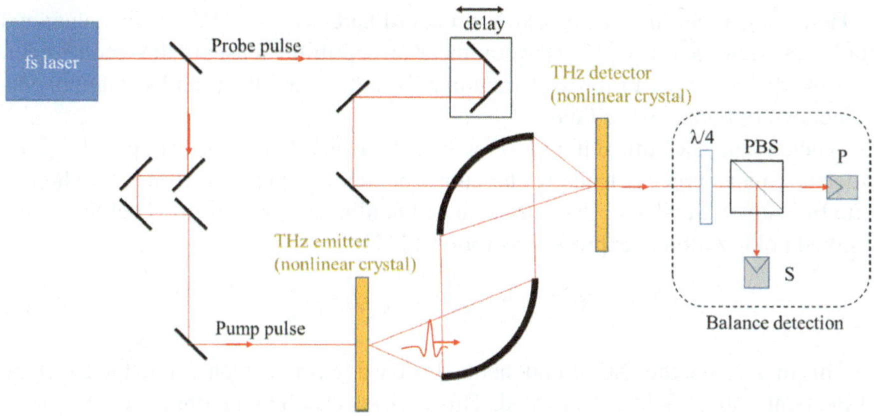

Fig. 2.7 Typical THz-TDS experimental setup for generating and detecting the THz electric field waveform by nonlinear crystals

as the emitter/detector of the THz waves, but most commonly the nonlinear optical processes such as optical rectification and Pockels effect in $\chi^{(2)}$ nonlinear crystals (ZnTe, GaP and $LiNbO_3$, for example) are used. Experimental setup used to generate and detect the THz waves are schematically illustrated in Fig. 2.7. It is based on pump-probe scheme, where the pump pulse (here, the electric field of emitted THz pulse) induces transient change of optical properties in matter and the subsequently incident gate pulse with a proper time delay probes its value. In THz-TDS, THz waves generated in the (emitter) nonlinear crystal is focused onto another (detector) crystal. The electric field of THz waves induces a transient birefringence in the detector crystal through Pockels effect. This birefringence reflects the transient amplitude of incident THz electric field. Therefore, by detecting the change of polarization in transmitted probe pulse induced by this birefringence while scanning time delay between the THz and probe pulse, temporal waveform of THz field is measured. Detailed explanations of the generation and detection methods are given in the following subsections.

2.2.2 *Generation of THz Waves via Optical Rectification*

Generation of THz waves is commonly realized by using optical rectification process in second-order nonlinear optical crystals. In a non-inversion symmetric media, second nonlinear optical coefficient is allowed to exist. General expression of such process is represented as follows:

$$P_i^{(2)} = \varepsilon_0 \sum_{j,k} \chi_{ijk}^{(2)} E_j E_k \tag{2.13}$$

Here, (i, j, k) are indices of Cartesian coordinates (x, y, z). $P_i^{(2)}$ is the generated nonlinear polarization, $\chi_{ijk}^{(2)}$ is the nonlinear susceptibility tensor of the second order, ε_0 is the dielectric permittivity of vacuum, E_j and E_k are the j- and k- components of incident optical electric field.

When an intense optical field $E(t) = E_0\, e^{-i\omega t}$ is incident on the sample, this 2nd–order nonlinear process leads to the sum frequency generation from two photons with frequency ω and $-\omega$. Using the reduced nonlinear susceptibility coefficient, the resulted polarization is expressed as follows [17]:

$$P^{(2)} = 2\varepsilon_0 \chi^{(2)}(0, \omega, -\omega) E_0^2 \tag{2.14}$$

This means that the D.C. component whose waveform is identical to the envelope of incident optical field is generated. This process is called the optical rectification. Usually, femtosecond pulses that has temporal width of ~100 fs are used for THz-TDS. Therefore, the THz pulse generated from this polarization has a duration period comparable to the incident optical field, typically several hundred femtoseconds to picosecond.

The kind of nonlinear crystals used for THz generation is chosen considering the phase matching condition between THz and optical pulses. In the collinear generation setup at 800 nm optical wavelength, the most commonly used nonlinear crystal is the ZnTe, because it has a relatively large nonlinear susceptibility ($\chi_{123}^{(2)}$ ~8 pm/V) and also the refractive index for 800 nm and THz frequencies are almost the same [n_{opt} (812 nm) = n_{THz} (1.69 THz)] and therefore, exhibits high conversion efficiency from optical- to the THz fields. Various other crystals are used depending on the application.

2.2.3 *Generation of Intense THz Waves by Tilted Wave Front Technique*

It is known that there exists other nonlinear optical crystals that exhibits much larger values of nonlinear susceptibility than ZnTe. For example, that of $LiNbO_3$ is known to be several times larger ($\chi_{333}^{(2)}$ ~27 pm/V). However, the large refractive index mismatch between THz (n_{THz} ~5.2) and optical frequencies ($n^{gr}{}_{\mathrm{opt}}$ ~2.3) does not allow this crystal to be used in THz generation in the collinear configuration. Note that n_{opt} ~2.3 is smaller than n_{THz} ~5.2, which means that the group velocity of optical pulse $v^{gr}{}_{opt}$ inside the $LiNbO_3$ crystal is around 2 times faster than the phase velocity v_{THz} of generated THz pulse. Therefore, if a usual pump pulse with its wave front perpendicular to the wave vector was used, the generated THz pulses cannot be coherently added up at every position in the crystal due to the phase mismatch.

In order to solve this phase-mismatching problem, a novel method known as tilted wave front technique was proposed and demonstrated by the groups of Hebling [18]

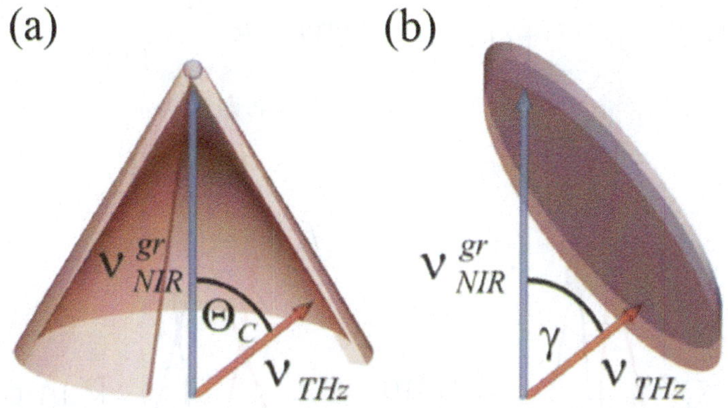

Fig. 2.8 THz generation based on nonlinear Cherenkov phase matching with **a** a well-focused point-like optical pulse and **b** plane wave with tilted wave front [19]. Figure reprinted with permission from [19], © 2008 OSA

and Nelson [19]. Figure 2.8 schematically illustrates its principle. When a well-focused, point-like fs pulse is incident on the $LiNbO_3$ crystal instead of plane-wave (a), the THz pulse generated by optical rectification will be radiated toward the direction away from the original propagation direction of fs pulse. This angle γ is given by:

$$\cos\gamma = \frac{v_{THz}}{v^{gr}_{opt.}} = \frac{n^{gr}_{opt.}}{n_{THz}}, \tag{2.15}$$

where $n^{gr}{}_{opt.}$ is the group refractive index of the material in the optical frequency.

This relation is determined from the condition that the phase velocity of THz and group velocity of optical pulse is fulfilled along the propagation direction of generated THz pulse. The idea of the method is to use the incident optical pulse that has a wave front tilted from the original direction by this amount of angle [Fig. 2.8b], so that the phase of generated THz pulse at different positions are always the same throughout the propagation of incident pulse in the crystal. In other words, the radiated THz wave will get coherently amplified during the propagation of light pulse within the $LiNbO_3$ crystal. Due to its resemblance to Cherenkov radiation of ultraviolet light from dumped electron motion, this method is also often called Cherenkov phase matching technique.

In order to obtain the tilted wave front, diffraction grating is used [Fig. 2.9a]. After the diffraction, optical pulse is tightly focused on the $LiNbO_3$ crystal by using a cylindrical lens to form a line-shaped beam. Generated THz waves are collected and collimated by a parabolic mirror.

With this method, quite intense THz electric field with its peak amplitude reaching several hundreds of kV/cm can be obtained using regenerative-amplified 800 nm pump pulses of ~100 fs time width, repetition rate of 1 kHz and pulse energy of several mJ. However, it was revealed after the development of this technique that

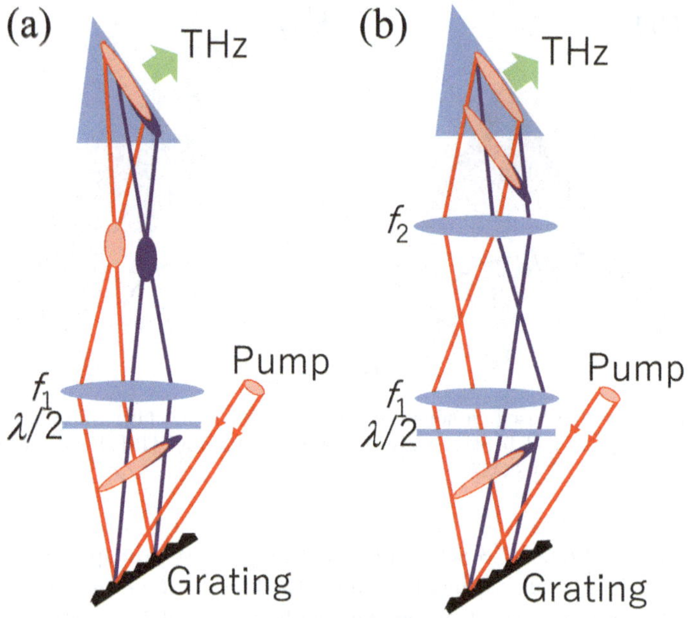

Fig. 2.9 Experimental setup used for Cherenkov phase matching in $LiNbO_3$ using **a** a grating and a focusing lens and **b** a grating and 4f-lens pair. In (**a**), the tilt angle of the wavefront changes around the focus point and therefore, the operation requires exact alignment of the focus system. In (**b**), the tilt angle around the focus remains almost constant, thus a more stable operation than (**a**) can be accomplished

in the imaging system consisting of grating, $LiNbO_3$ crystal and only one lens, slight errors of focus system can cause a great change of tilting angle, resulting in the distortion of wave front and weaker peak amplitude of the output THz pulse waveform, because the tilt angle of the wave front strongly changes near the focus point. In order to overcome this problem, Hirori et al. [20] employed a novel system which is known as 4f-lens configuration as shown in Fig. 2.9b. In this method, a pair of cylindrical lenses are placed between the grating and $LiNbO_3$ crystal. By using such imaging system, the image of the spot on the grating is directly transferred onto the $LiNbO_3$ crystal while maintaining the tilt angle of the wave front near the focus point. This method enables more stable operation than imaging with only one lens, because around the $LiNbO_3$ crystal the beam is almost collimated, and therefore, the tilt angle of the wave front does not differ significantly for small alignment errors. With this method, they obtained the intense THz pulse that reached ~1 MV/cm. As will be shown in Chap. 3, the measurement in this Thesis is based on similar setup, and THz fields with peak electric field of several hundred kV/cm (magnetic field of ~0.1 T order) are obtained.

2.2.4 Detection of Magnetization Dynamics by Magnetooptical (MO) Effects

In this study, the magnetization dynamics of $ErFeO_3$ induced by the THz magnetic fields are detected by means of magnetooptical Faraday effect. When light transmits through a transparent medium magnetized along the direction parallel to the wave vector of light, it feels different refractive indices depending on its helicity. As the linearly polarized light can be decomposed to right- and left- circularly polarized components, such difference leads to the rotation of polarization direction after transmission. This phenomenon is called magnetooptical Faraday effect.

Phenomenologically, the dielectric permittivity of such medium is represented as follows [21]:

$$\varepsilon_{ij} = \begin{pmatrix} \varepsilon_{xx} & \varepsilon_{xy} & 0 \\ -\varepsilon_{xy} & \varepsilon_{yy} & 0 \\ 0 & 0 & \varepsilon_{zz} \end{pmatrix} \tag{2.16}$$

In the following we assume $\varepsilon_{xx} = \varepsilon_{yy}$ for simplicity (no birefringence case). We can assume that each components can be expanded into the Taylor series of magnetization $\boldsymbol{M}$. On the other hand, according to the Onsager's relations, each components are connected to its diagonal counterparts through following equation:

$$\varepsilon_{ij}(M) = \varepsilon_{ji}(-M) \tag{2.17}$$

This tells us that the diagonal elements are even and non-diagonal elements are odd with respect to M. Therefore, we can assume $\varepsilon_{xy}(M) = \varepsilon_{xy}^{(1)} \cdot M$ as the first approximation. For optical field propagating along the z direction with frequency ω, there exist two eigenfunctions for refractive index $n_\pm$ and the corresponding normal modes $E_\pm$:

$$\begin{cases} n_\pm^2 = \varepsilon_{xx} \pm i\varepsilon_{xy}, \\ \boldsymbol{E}_\pm = E_0(\hat{\boldsymbol{x}} \pm i\hat{\boldsymbol{y}})e^{-i\omega(t-\frac{n_\pm}{c}z)} \end{cases} \tag{2.18}$$

The normal modes shown above correspond to either right- (+) or left-circular (−) polarizations. Using these, the Faraday rotation angle for a linearly polarized light propagating through the medium by a length L can be expressed as follows:

$$\theta_F = \frac{-\omega\varepsilon_{xy}'' L}{2cn} = \frac{-\omega\varepsilon_{xy}^{(1)''} L}{2cn} M \tag{2.19}$$

It can be seen that the Faraday rotation angle is proportional to the sample thickness and magnetization. Therefore, the measurement of transient Faraday rotation allows us to directly detect the wavevector- component of the spin dynamics. The coefficient for Faraday rotation against magnetization in a unit length is determined intrinsically by the material property, and is called Verdet constant.

In a more microscopic view, magnetooptical effect is explained in the following manner [Fig. 2.10a]. Consider the optical transition between two-level system, where the ground state |g> has a singlet orbital state with $L = 0$ and the excited state |e> has $L = 1$. If the medium was placed under magnetic field (either external or internal field such as exchange field), it results in the lifting of degeneracy about spin states. This alone does not affect the optical transitions because the optical field does not directly interacts with the spin. However, when spin–orbit coupling is present, it causes the $L = 1$ states to split into $M_z = +1$ and -1 states ($M_z = 0$ is neglected, since it does not affect magnetooptical effects). Optical transition from the ground state into these states are allowed with circularly polarized light with opposite helicities: This means that the transition frequency is shifted depending on their helicities. Therefore, it results in the shift of the corresponding absorption peaks and hence, difference in refractive indices below the transition frequency. From detailed analysis using the Kubo formula, the following relation is derived:

$$\varepsilon''_{xy} = \frac{-Ne^2 f_0 \Delta_{so}}{4m\varepsilon_0\omega} \cdot \frac{(\omega - \omega_0)^2 - \gamma^2}{\left\{(\omega - \omega_0)^2 + \gamma^2\right\}^2}, \tag{2.20}$$

where Δ_{so} is the spin–orbit coupling energy, m is the effective mass of electron, f_0 is the oscillator strength, N is the number of oscillators, e is electron charge and γ is the line width. As shown in Fig. 2.10b, $\varepsilon''_{xy}\varepsilon_{xy}$" increases near the optical transition frequencies. The detailed spectral shapes of magnetooptical coefficients in a realistic medium vary strongly depending on the kinds of transitions, from which many useful information on the electronic energy levels can be extracted.

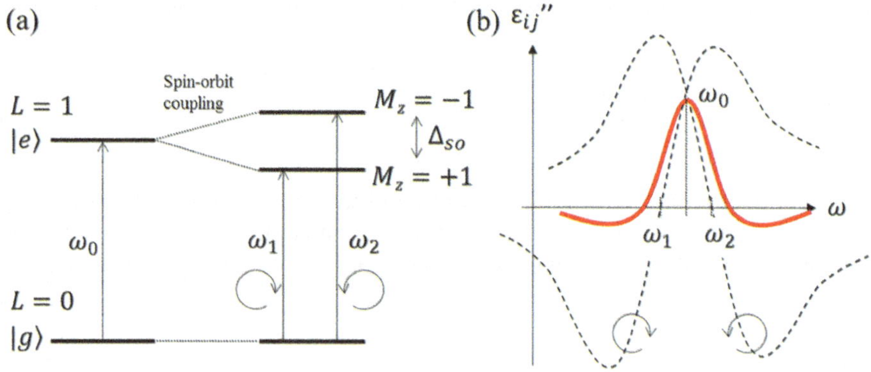

Fig. 2.10 **a** Origin of magnetooptical effects due to spin–orbit coupling. **b** Diagonal component of the dielectric permittivity caused by the shift of transition frequencies depending on the polarization of incident light

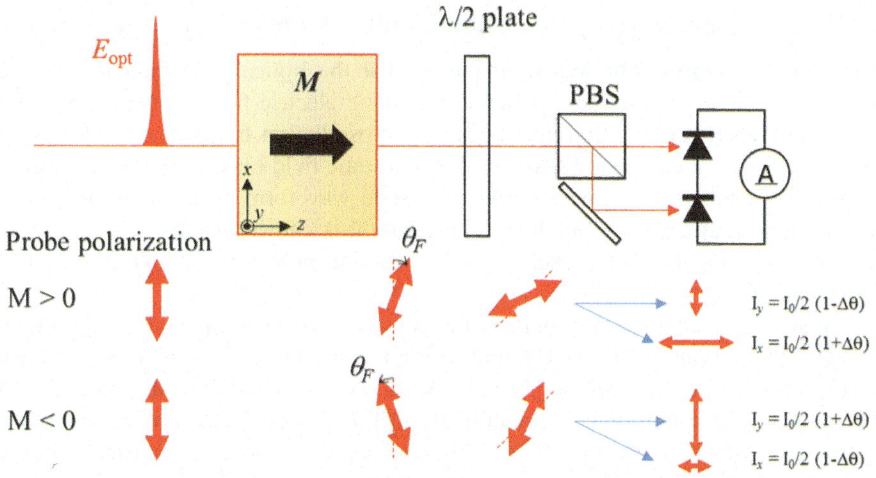

Fig. 2.11 Measurement of the angle of Faraday rotation by balance detection using half-wave plate and polarizing beam splitter (PBS). The arrows indicate the polarization states of the probe light within the *x*–*y* plane

In order to measure the Faraday rotation, experiment setup shown in Fig. 2.11 is used in this study. Here, the incident light is linearly polarized. The angle of half-wave plate (HWP) is originally set so that the output polarization is directed towards 45 degree without magnetization and therefore, the difference of the intensities of horizontal and vertical components is zero. When the sample is magnetized, on the other hand, the light output from sample will be slightly rotated by θ_F due to Faraday rotation, leading to the unbalance of horizontal and vertical intensities. From this, the angle of Faraday rotation can be deduced.

2.2.5 Detection of THz Waves by Electro-Optic (EO) Sampling

THz waves generated in the above-mentioned nonlinear optical processes can be detected by using again the second-order nonlinear optical process. This time, we consider the case where the electric fields of optical (E_O) and THz (E_{THz}) pulses are simultaneously incident on the crystal. The nonlinear polarization can be expressed by:

$$P_i^{(2)}(\omega) = \varepsilon_0 \sum_{j,k} \chi_{ijk}^{(2)}(\omega, \omega, 0) E_j^O(\omega) E_k^{THz}(0) \equiv \varepsilon_0 \sum_j \chi_{ij}^{(2)}(\omega) E_j^O(\omega). \quad (2.21)$$

Here, $\chi_{ij}^{(2)}(\omega) \equiv \sum_k \chi_{ijk}^{(2)}(\omega, \omega, 0) E_k^{THz}(0)$ describes the THz-field induced susceptibility tensor. This equation means that the optical birefringence can be induced in the crystal by the application of static electric field. This phenomenon is called Pockels effect. Note that because the oscillation frequency of THz pulse is much slower, it can be regarded as almost a static field bias for the optical pulse. This effect enables us to measure the THz field waveform by detecting the value of transient birefringence. In a lossless medium it is known that $\chi_{ijk}{}^{(2)}(\omega, \omega, 0) = \chi_{ijk}{}^{(2)}(0, \omega, -\omega)$, thus the Pockels effect has the same nonlinear coefficient as optical rectification.

Figure 2.12 schematically depicts the conventional experimental configuration used for the measurement of THz fields using Pockels effect. This method is known as electro-optic (EO) sampling [17, 22, 23]. In order to satisfy the abovementioned phase-matching condition between optical and THz pulses, ZnTe (110) or GaP (110) are often used for detection. (ZnTe exhibits a larger sensitivity to be used with the 800 nm optical wavelength, but its detectable THz bandwidth is limited by absorption due to optical phonon in the higher frequency region (approximately > 2 THz). GaP shows somewhat less sensitivity with 800 nm pulse but exhibits a broader sensitivity, reaching up to ~7 THz.) Then, substituting

$$E_O = \frac{E_O}{\sqrt{2}} \begin{pmatrix} 1 \\ -1 \\ 0 \end{pmatrix} \text{ and } E_{THz} = \frac{E_{THz}}{\sqrt{2}} \begin{pmatrix} 1 \\ -1 \\ 0 \end{pmatrix} \tag{2.22}$$

into Eq. 2.21, the resulting polarization can be represented as follows:

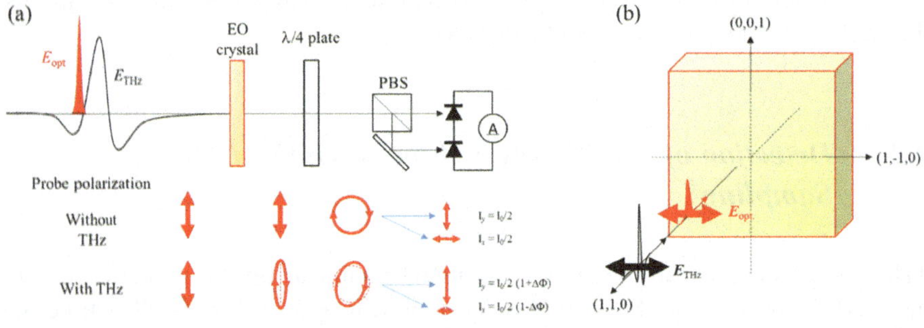

Fig. 2.12 Measurement of the THz electric field with EO sampling method. **b** Configuration of incident polarization and the crystal axis

$$\begin{pmatrix} P_x \\ P_y \\ P_z \end{pmatrix} = 4\varepsilon_0 d_{14} \begin{pmatrix} 0\,0\,0\,1\,0\,0 \\ 0\,0\,0\,0\,1\,0 \\ 0\,0\,0\,0\,0\,1 \end{pmatrix} \begin{pmatrix} E_{O,x}E_{THz,x} \\ E_{O,y}E_{THz,y} \\ E_{O,z}E_{THz,z} \\ E_{O,y}E_{THz,z} + E_{O,z}E_{THz,y} \\ E_{O,z}E_{THz,x} + E_{O,x}E_{THz,z} \\ E_{O,x}E_{THz,y} + E_{O,y}E_{THz,x} \end{pmatrix}$$

$$= -4\varepsilon_0 d_{14} E_O E_{THz} e_z (\perp E_O). \tag{2.23}$$

Here, $d_{14} = (1/2)\ \chi_{123}{}^{(2)}$. As can be seen, the polarization generated by Pockels effect is proportional to the THz electric field amplitude, and is polarized perpendicular to the original optical field, indicating that the output optical field is elliptically polarized. Macroscopically, this can be interpreted as an appearance of birefringence towards 45° from the original direction of polarization. Then, after the propagation along the crystal thickness L, the phase difference between horizontal and vertical electric fields are represented as:

$$\Delta\phi = \left(n_y - n_x\right)\frac{\omega L}{c} = n_{opt.}^3 \frac{\omega L}{c} d_{14} E_{THz}. \tag{2.24}$$

Using quarter wave plate (QWP) with its optical axis fixed horizontally, the optical field having such phase difference is transformed into an almost circular, but slightly elliptical polarization after QWP. Difference of the intensities of horizontal and vertical polarizations measured by using polarizing beam splitter (PBS) and balanced photodetector is given by:

$$\Delta I = I_y - I_x = I_O \Delta\phi = n_{opt.}^3 \frac{I_O \omega L}{c} d_{14} E_{THz} \propto E_{THz}, \tag{2.25}$$

which is proportional to the incident THz field strength. Therefore, with EO sampling method, we can evaluate the absolute amplitude of THz electric field as long the values of nonlinear coefficients of the EO crystals are known.

2.3 THz Magnetic Nearfields in Split-Ring Resonator (SRR)

In this section the electromagnetic properties of the metallic SRR structure is overviewed, which is the key device used to excite spin resonance in $ErFeO_3$ in this study. Also, some previous reports that focused on the usage of SRR-nearfields are reviewed.

2.3.1 *Electric Excitation of Magnetic Resonance Mode in SRR*

As mentioned in the introduction, SRR is one of the most frequently used structures to realize metamaterial, that is, artificial subwavelength structures that holds unnaturally large magnetic response for the electromagnetic waves. Since its invention, various types of similar structures have been invented. Some examples of the SRR structures are shown in Fig. 2.13 [24]. In conventional materials, the electric polarization is generated only from electric field ($\boldsymbol{D} = \varepsilon \boldsymbol{E}$) and magnetization change is induced only by magnetic field of electromagnetic waves ($\boldsymbol{B} = \mu \boldsymbol{H}$). However, in metamaterials the magnetic and electric polarizations inside the structure are strongly coupled with each other. More general form of above equations then becomes: $\boldsymbol{D} = \varepsilon \boldsymbol{E} + \alpha \boldsymbol{H}$ and $\boldsymbol{B} = \beta \boldsymbol{E} + \mu \boldsymbol{H}$, respectively. Such metamaterials are called "bianisotropic" metamaterials.

In the case of SRR (a) in Fig. 2.13, for example, if the incident THz electric field was applied parallel to the gap of the ring, the incident field generates not only the electric polarization in the gap part but also the oscillating magnetic field in the ring due to the circulating current. Conversely, if we apply the incident field such that $\boldsymbol{H}_{\mathrm{THz}} \perp$ ring, then the magnetic field induces the electric field in the gap part.

The strength of such interconversion between magnetic and electric components depends on the structure designs and it becomes less pronounced for isotropic structures. In a more isotropic SRR shown in (b), for example, if it was excited by the incident THz magnetic field penetrating the ring, then the direction of the electric dipole moments induced in the gaps of the inner and outer rings are opposite. Therefore, as a whole it exhibits smaller electric dipole moment. In a completely symmetric structure (c), the electric-dipole moments induced in the circular current mode are completely cancelled and only the pure magnetic response can be observed, i.e., $\alpha = \beta = 0$.

Abovementioned fact that a strong magnetic field can be created by using the electric field component of the incident light field in bianisotropic metamaterials makes it possible to regard them as a kind of transducer that converts electric energy to the magnetic one in the nearfield region. Also, such electric-field induced magnetic field can sometimes exceed the original magnetic field contained in the incident THz

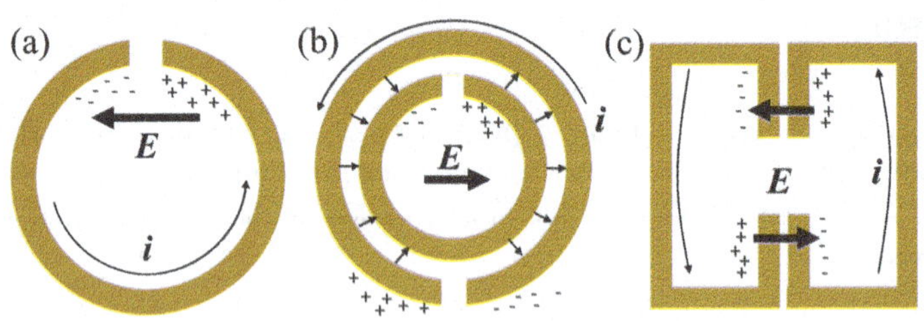

Fig. 2.13 Examples of the SRR structures. **a** Single-turn SRR, **b** dual-SRR and **c** symmetric dual-SRR

pulse by orders of magnitude. In the following Chaps. 3 and 4, we use this feature to realize a strongly enhanced spin precession in SRR structures.

2.3.2 Usage of Field Enhancement Effects in THz-SRR

As the circulating-current resonance mode of SRR can be viewed as an *LC* resonance, during the resonance, there exist the moments wherein all of the electromagnetic energy is confined either in the capacitance (gap) or inductance (ring). This leads to the appearance of strongly localized and enhanced electric field in the gap or magnetic field near the corner of the ring. In terms of plasmonics, such positions are called "hot spot". As the energy is strongly confined in such regions, the resonance mode is highly sensitive to the change of dielectric properties in these positions. Such feature is attractive for sensing applications [25]. Also, the high electric-field enhancement factor that can reach 10^1 ~10^2 in the hot spot provides useful platform for nonlinear phenomena. Liu et al. [26] reported that the enhanced electric field of intense THz waves in SRR can even cause destructive phenomenon in a metal–insulator transition material, VO_2.

In contrast to the successful application of electric-field enhancement in THz SRRs, the utilization of the magnetic-field enhancement effect in SRR on the dynamical control of condensed matter system in the THz region are quite limited. Moreover, due to the fact that the detection of magnetic field of electromagnetic waves is much more difficult than electric field, there are only a few reports which directly visualized such enhanced magnetic nearfield of the SRR. One of the few demonstrations was reported by Kumar et al. [27] in which they applied magnetooptical (MO) Faraday imaging to the SRR and directly visualized the THz magnetic nearfield created by the SRR [Fig. 2.14a]. They used a paramagnetic substrate (TGG: terbium gallium garnet) that exhibits high Verdet constant and ultrafast (ps) response time and fabricated the SRR structures on its surface. When the SRR was irradiated by the THz electric field, the oscillating current-induced magnetic field rotates the polarization of incident probe pulse. By raster-scanning the probe positions, they measured the spatial profile of the magnetic nearfield [Fig. 2.14b]. The result was is in good agreement with the result expected from electromagnetic calculation and suggested that the magnetic fields are possibly enhanced by orders of magnitude compared to the incident THz magnetic field near the corner of SRR.

2.3.3 Coupling of SRR and THz Spin Resonance Mediated by Magnetic Nearfield

Although the THz-SRR magnetic nearfield has been utilized in only a few number of demonstrations at the moment, it has the potential to greatly extend the scope of THz

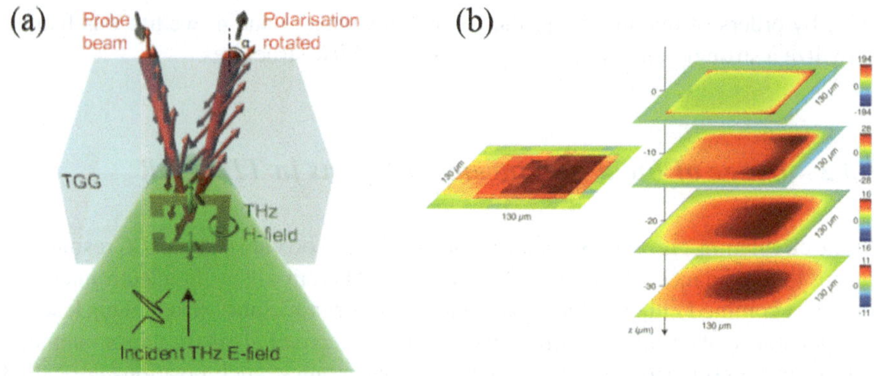

Fig. 2.14 **a** Schematic of the magnetooptical (MO) imaging with TGG crystal [27]. **b** Spatial profile of the out-of-plane magnetic nearfield H_z around SRR structure measured by MO imaging (left) and calculated by electromagnetic simulation with finite-difference-time-domain (FDTD) method. Figure reprinted with permission from [27], © 2012 OSA

magnetic control of spin system. Firstly, the conversion of incident THz electric field into the highly enhanced magnetic nearfield mentioned above enables the intense magnetic field to be generated under the incident THz field with smaller amplitude, which is a desirable feature for inducing strong magnetic-field perturbation on the spin system while avoiding the THz electric field effects. Secondly, the strong localization of magnetic field energy around the SRR may enable not only an efficient "excitation" of spin resonance but also a more sophisticated dynamics, mediated by the interactive energy transfer between the SRR and the spin system.

Schneider et al. [28] reported that the coupling between magnetic-dipole resonance mode of SRR structure and the THz electron spin resonance (ESR) of magnetic materials actually exists. They fabricated the array of bianisotropic SRRs on a paramagnetic (gadolinium gallium garnet: GGG) substrate. A strong static magnetic field was applied on the substrate in a direction parallel to the surface, in order to tune the spin resonance frequency into the sub THz region. Through transmission measurement of such SRR array fabricated on magnetically biased spin system using backward-wave oscillator (BWO) as the continuous-wave (CW) THz source, they showed that the absorption peaks corresponding to the SRR and spin resonance of the substrate exhibit anti-crossing behavior (Fig. 2.15). They concluded that such behavior was caused by the coupling between the spin and SRR resonances mediated by the magnetic nearfield.

The coupling of electronic transitions with the external cavity modes has been extensively studied in the optical frequency regions in relation to the study fields such as cavity quantum electrodynamics. In microwave regions, coupling of magnetic substance with the metallic antenna structures has been utilized to realize polarization-control devices. In the intermediate frequency region of THz, on the other hand, such coupling phenomena are yet thoroughly studied. Especially, the direct measurement of the *dynamics* of spin system under the strong influence of

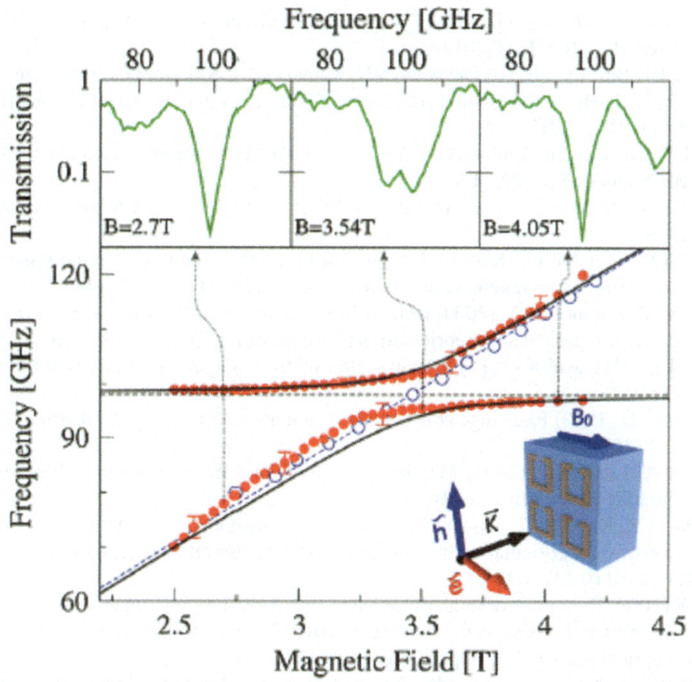

Fig. 2.15 Anticrossing of SRR and spin resonance mediated by magnetic nearfield coupling observed in CW millimeter-wave transmission measurements [28]. Figure reprinted with permission from [28], © 2009 AIP

magnetic cavity (SRR resonance mode in this case) have not been performed. To clarify this, in the subsequent Chap. 3 we examine the behavior of such spin dynamics in the time-domain using THz-TDS.

References

1. White RL (1969) Review of recent work on the magnetic and spectroscopic properties of the rare-earth orthoferrites. J Appl Phys 40:1061
2. Bertaut EF, Mareschal J (1967) Etude de la structure magnetique des chromites d'erbium et de neodyme par diffraction neutronique. Solid Stat Commun 5:93
3. Dzyaloshinskii IE (1957) Thermodynamic theory of weak ferromagnetism in antiferromagnetic substances. Sov Phys JETP 5:1259
4. Moriya T (1960) Anisotropic superexchange interaction and weak ferromagnetism. Phys Rev 120:91
5. Herrmann GF (1964) Magnetic resonances and susceptibility in orthoferrites. Phys Rev 133:A1334
6. Herrmann GF (1963) Resonance and high frequency susceptibility in canted antiferromagnetic substances. J Phys Chem Solids 24:597

7. Shapiro SM, Axe JD, Remeika JP (1974) Neutron-scattering studies of spin waves in rare-earth orthoferrites. Phys Rev B 10:2014
8. Tsymbal LT, Bazaliy YB, Derkachenko VN, Kamenev VI, Kakazei GN, Palomares FJ, Wigen PE (2007) Magnetic and structural properties of spin-reorientation transitions in orthoferrites. J Appl Phys 101:123919
9. Belov KP, Zvezdin AK, Kadomtseva AM, Levitin RZ (1976) Spin-reorientation transitions in rare-earth magnets. Sov Phys Usp 19:7
10. Morin FJ (1950) Magnetic susceptibility of AFe_2O_3 and AFe_2O_3 with added titanium. Phys Rev 78:819
11. Bazaliy YB, Tsymbal LT, Kakazei GN, Wigen PE (2004) The role of erbium magnetization anisotropy during the magnetic reorientation transition in $ErFeO_3$. J Appl Phys 95:6622
12. Belyaeva AI, Baranova KV (2009) Role of Er3+ ion in the formation of the $ErFeO_3$ magnetic properties in the region of spin-reorientation phase transition. Bull Russ Acad Sci Phys 73:1056
13. Koshizuka N, Hayashi K (1988) Raman scattering from magnon excitations in $RFeO_3$. J Phys Soc Jpn 57:4418
14. Gorodetsky G (1960) Exchange constants in orthoferrites $YFeO_3$ and $LuFeO_3$. J Phys Chem Solids 30:1745
15. Perroni CA, Liebsch A (2006) Magnetization dynamics in dysprosium orthoferrites via the inverse Faraday effect. Phys Rev B 74:134430
16. Yamaguchi K, Kurihara T, Minami Y, Nakajima M, Suemoto T (2013) Terahertz time-domain observation of spin reorientation in orthoferrite $ErFeO_3$ through magnetic free induction decay. Phys Rev Lett 110:137204
17. Lee Y-S (2009) Principles of terahertz science and technology. Springer
18. Fülöp JA, Pálfalvi L, Almási G, Hebling J (2010) Design of high-energy terahertz sources based on optical rectification. Opt Express 18:12311
19. Hebling J, Yeh K-L, Hoffmann MC, Bartal B, Nelson KA (2008) Generation of high-power terahertz pulses by tilted-pulse-front excitation and their application possibilities. J Opt Soc Am B 25:B6
20. Hirori H, Doi A, Blanchard F, Tanaka K (2011) Single-cycle terahertz pulses with amplitudes exceeding 1 mv/cm generated by optical rectification in $LiNbO_3$. Appl Phys Lett 98:091106
21. Sato K. Light and Magnetism
22. Nahata A, Auston DH, Heinz TF, Wu C (1996) Coherent detection of freely propagating terahertz radiation by electro-optic sampling. Appl Phys Lett 68:150
23. Nahata A, Weling AS, Heinz TF (1996) A wideband coherent terahertz spectroscopy system using optical rectification and electro-optic sampling. Appl Phys Lett 69:2321
24. Katsarakis N, Koschny T, Kafesaki M, Economou EN, Soukoulis CM (2004) Electric coupling to the magnetic resonance of split ring resonators. Appl Phys Lett 84:15
25. Willets KA, Van Duyne RP (2007) Localized surface plasmon resonance spectroscopy and sensing. Annu Rev Phys Chem 58:267
26. Liu M, Hwang HY, Tao H, Strikwerda AC, Fan K, Keiser GR, Sternbach AJ, West KG, Kittiwatanakul S, Lu J, Wolf SA, Omenetto FG, Zhang X, Nelson KA, Averitt RD (2012) Terahertz-field-induced insulator-to-metal transition in vanadium dioxide metamaterial. Nature 487:345
27. Kumar N, Strikwerda AC, Fan K, Zhang X, Averitt RD, Planken PCM, Adam AJL (2012) THz near-field Faraday imaging in hybrid metamaterials. Opt Exp 20(10):11277
28. Schneider A, Shuvaev A, Engelbrecht S, Demokritov SO, Pimenov A (2009) Electrically excited inverse electron spin resonance in a split-ring metamaterial resonator. Phys Rev Lett 103:103907

Chapter 3
Resonant Enhancement of Spin Precession by SRR-induced Magnetic Nearfields and Interactive Energy Transfer

Abstract It has been shown in Chap. 2 that split ring resonator (SRR) can create a strongly enhanced magnetic nearfield in its vicinity. By using such locally enhanced terahertz magnetic field, we can expect to greatly increase the amplitude of the spin precession. The main scopes of this chapter are twofold: (1) to demonstrate the excitation of spin precession by using the enhanced magnetic nearfield of SRR, and (2) to investigate the dynamical behavior of spin system in the time domain. For this purpose, in this chapter we first describe the single crystal sample of $ErFeO_3$, on which the SRR structures are fabricated (Sect. 3.1). We perform the intense THz pump-optical Faraday probe experiment (Sect. 3.2) to measure the dynamics of spin system under the influence of SRR, aiming at revealing the coupled-resonance behavior between the spin precession and the electromagnetic fields in SRR (Sects. 3.3 and 3.4).

Keywords $ErFeO_3$ · Floating zone growth · Coupled resonance · Finite-difference time-domain simulation

3.1 Sample Fabrication

The samples used in the study are the $ErFeO_3$ single crystals and the SRR structures. In this section, their fabrication procedures are summarized.

3.1.1 Fabrication of Single-Crystal $ErFeO_3$ (001) Plate with Floating-Zone Method

The sample used for the experiment is the (001)-plane cut crystal of $ErFeO_3$, grown by Floating-zone (FZ) method [123]. In FZ, intense light of halogen lamps are focused by ellipsoidal mirrors on the compressed rods of row material powders and melt them. The melting zone is held between the seed and feed rods during the growth by surface tension of liquid phase. By scanning the rods downwards the melting zone moves upward, leaving a single crystal behind. Here, four halogen lamps with 500 W

T. Kurihara, *Observation and Control of Magnetic Order Dynamics by Terahertz Magnetic Nearfield*, Springer Theses, https://doi.org/10.1007/978-981-16-8793-8_3

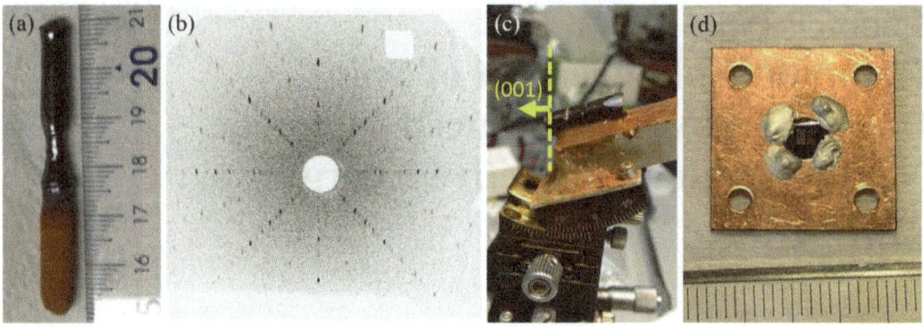

Fig. 3.1 **a** Single crystal rod of the FZ-grown $ErFeO_3$. **b** X-ray Laue pattern of the crystal, indicating *c*-plane. **c** Cutting of the crystal rod in a *c*-plane-normal direction. **d** Photograph of the resulted sample mounted on a phosphor-bronze cryostat holder

output power, the growth rate of 4 mm/h, O_2 gas flow rate of ~1 L/h were used. As the row powders, the $ErFeO_3$ purchased from Furuuchi Chemical Corporation were used. In order to adjust the stoichiometry (molar ratio) of Er and Fe in the crystal, small amount of Fe_3O_4 powder were added into the $ErFeO_3$ powder. The powder was compressed in a rod form with water pressure of approximately 400 kgf/cm^2, sintered in a muffle furnace at ~1100 °C for 10–12 h and then sintered again in the FZ furnace below the melting temperature before the growth. The FZ growth procedure was conducted in the Materials Synthesis Section in the ISSP, University of Tokyo.

The obtained single crystal rod of $ErFeO_3$ after growth is shown in Fig. 3.1a. Its orientation direction was measured by using X-ray Laue diffraction [Fig. 3.1b], then cut perpendicular to the *c*(001)-axis of the crystal [Fig. 3.1c]. It was then mechanically polished to a thickness of approximately 130 μm, which was thin enough for the 800 nm wavelength to transmit. The sample was mounted on a copper plate, which was then set on a cryostat (Oxford Microstat) during the measurement [Fig. 3.1(d)].

3.1.2 Optical Characterization of $ErFeO_3$ Sample

As characterization, we measured the optical absorption of the fabricated $ErFeO_3$ (001) sample. Here, a UV–VIS-IR spectrophotometer V-570 of JASCO Corporation was used in the transmission measurement mode. Figure 3.2a shows the obtained spectra in the wavelength region from 500 to 2000 nm. The absorption in the visible region is dominated by the strong charge transfer transition from legand (O^{2-}) to the Fe^{3+} ion (short wavelength region [$\lambda < 600$ nm]) and d-d transition peaks within the Fe^{3+} ion (peaks centered at ~700 and ~1000 nm). In the IR region of around 1550 nm, we can see a smaller peak that can be ascribed to the f-f transition in Er^{3+} ion. These spectral features agree well with the previously reported spectra of $ErFeO_3$ single crystal [Fig. 3.2b] [1]. The probe wavelength of 800 nm coincides the dip of the two

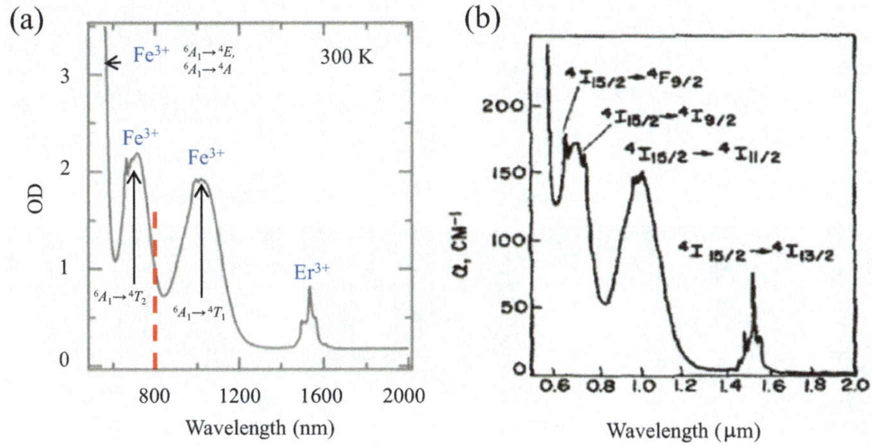

Fig. 3.2 **a** Absorption spectra of the FZ-grown $ErFeO_3$ (001) sample measured at room temperature. **b** Spectra for the same condition reported in previous study [1]. Figure reprinted with permission from [1]. © 1970 AIP [1]

d-d transitions, which allows us to conduct the magneto-optical measurement in the transmission regime, i.e., Faraday rotation to probe the spin dynamics.

Furthermore, we also conducted the microscopic Faraday imaging of the sample in order to visualize the magnetic domain structure, as shown in Fig. 3.3. Here we measured the Faraday image using the experimental setup described later (Fig. 3.8). The linearly polarized ($\boldsymbol{E}_{\mathrm{probe}}//a$-axis), collimated optical pulse with 800 nm wavelength incident on the sample was transferred through the objective lens onto the charge-coupled device (CCD) camera. Just before the CCD, a polarizer was placed, making an angle of 90° from the original direction of polarization. Note that the Faraday rotation is sensitive to the out-of-plane component of magnetization, so the perpendicular-magnetization domains can be visualized in this configuration.

Figure 3.3 shows the microscopic images of the domain structure measured at various temperature. In the high temperature (HT) region $T > T_H$ (120, 110 and 100 K), a maze structure composed of black and white regions of several 100 μm scale can be observed. The white and black regions correspond to the magnetization M pointing upward and downward, respectively. As the temperature is lowered, the average domain size continuously decreases. Within the SRPT temperature region $T_H > T > T_L$, the domain structure exhibits more complex pattern, changing its shape strongly with temperature. At the same time, the contrast of black and white area also continuously decreases. In the low temperature (LT) region $T < T_L$, the domain structure almost completely disappears. This means that the out-of-plane magnetization has disappeared in the LT region, which indicates that the magnetization easy-axis has rotated from out-of-plane at HT region to in-plane at LT, i.e., spin reorientation. Also, the threshold temperatures of SRPT region $T_H \fallingdotseq 96$ K and $T_L \fallingdotseq 85$ K well agree the previously reported values for $ErFeO_3$ ($T_H \fallingdotseq 96$ K and $T_L \fallingdotseq 85$ K) [2, 3].

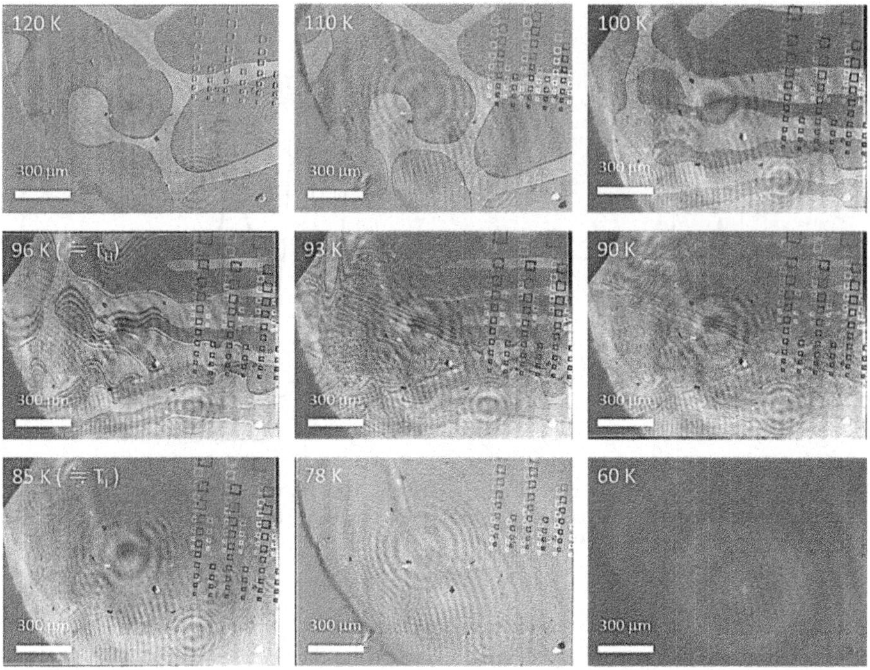

Fig. 3.3 Microscopic Faraday images measured around the SRPT temperature region. The concentric patterns in some images (see, e.g., 60 K) are the interference fringes due to misalignment of the focus system. The SRR arrays are visible on the right side of the image. Here, in order to eliminate the effect of non-uniform illumination caused by interference fringes, we subtracted the image of 60 K from each image. Contrast and brightness of the photos are also increased for visibility

It is known that the SRPT temperature is influenced strongly by the stoichiometry and dopants in the sample, which tend to lower the transition temperatures. The abovementioned agreement of $T_{H,L}$ then suggests that the fabricated sample is constituted of a pure $ErFeO_3$ single crystal with relatively few impurities. In the following experiments, the sample temperature is set at LT region, and the transient out-of-plane magnetization component during spin precession are detected by Faraday probe.

3.1.3 Fabrication of SRR Structures

The SRR structures were fabricated on the surface of $ErFeO_3$ using electron-beam (EB) lithography. The procedure is as follows: First, the resist was spin-coated on the substrate. It was then brought into the scanning electron microscope (SEM) chamber and the resist was irradiated by focused electron beam scanned in the shape of SRR. This makes the irradiated part solvable to the developer solution. The sample was then brought out of the SEM chamber and immersed in developer, which revealed the substrate surface formed in the shape of SRR. Next, it was set in evaporator and metallic thin films (Ni:Cr alloy of 5 nm thickness as buffer layer and the Al of approximately 200 nm thickness) were deposited. Lastly, the remaining part of the resist was removed by acetone, leaving the Al formed in the shape of SRR on the substrate. The fabrication process was conducted in collaboration with Quantum optics-Quantum transport group (Prof. Oto's group) in Chiba University.

Multiple SRR structures with different sizes (40–150 μm) were fabricated simultaneously on the same substrate in an attempt to tune resonance frequencies. Each elements were separated from each other by 1–2 times their lateral dimensions, in order to avoid the cross-coupling between them. Figure 3.4a shows the SRR structures fabricated on $ErFeO_3$ substrate. The SRR structure was fabricated on the substrate such that the gap of the structure is parallel to the *b*-axis of the $ErFeO_3$ crystal. During the measurement it was recognized that although each elements look similar under the optical microscope observation, unidentified factors such as a tiny scratch in the stripe of SRR, remnant of the photoresist or unremoved metal, etc. strongly influenced the response of individual SRR structures. Also, due to the non-uniformity of the $ErFeO_3$ substrate, the response of spin dynamics showed slight variation depending on the position in the sample. Therefore, in the following experiment the SRR which showed the largest amplitude of spin dynamics was selected. As the main element, we used the one shown in Fig. 3.4b which has the lateral dimension of 60×60 μm, line width of 10 μm and gap width of 10 μm, respectively.

The resonance frequencies of the SRR structures were numerically simulated using commercially available finite-difference time-domain (FDTD [4]) software (FullWAVE®). It is one of the standard methods for numerically calculating the time- and spatial evolution of the electromagnetic field and is frequently used for simulation of light field in the subwavelength region. As its name indicates, in FDTD, the Maxwell equations in the form of differential equations are solved step by step in the time domain. At each time step, the spatial distribution of electric and magnetic field vectors are calculated at every positions in the calculated region. The geometry used in the calculation is shown in Fig. 3.5. The out-of-plane magnetic field H_z

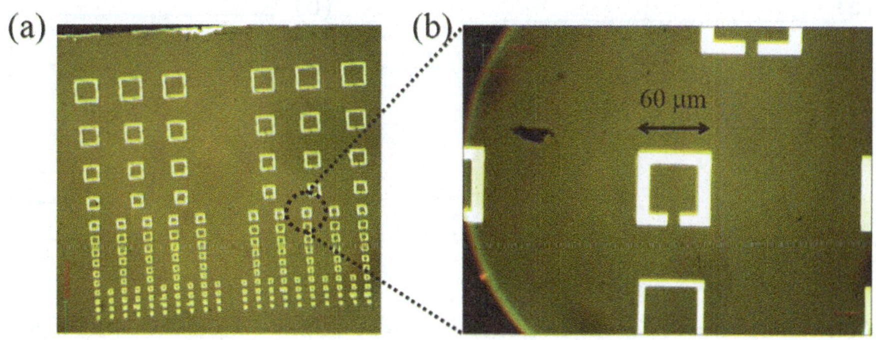

Fig. 3.4 **a** Array of SRR structures fabricated by electron-beam lithography. **b** Main SRR used in this chapter

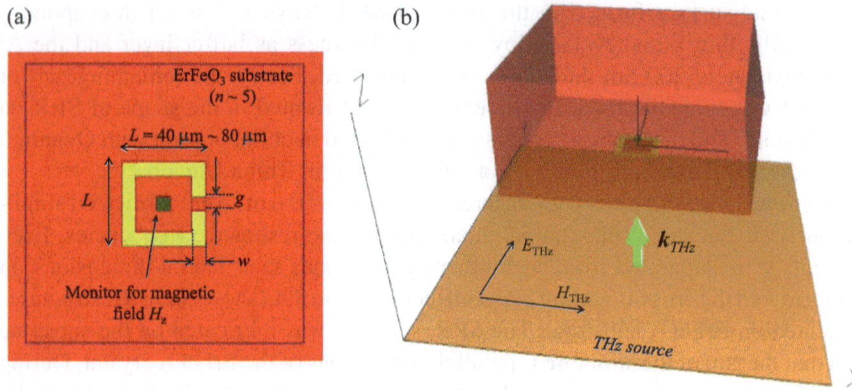

Fig. 3.5 **a** Calculation geometry of SRR resonance frequencies by FDTD method, viewed from (**a**) z- and **b** obliquely upward direction. Here, the width (w) and gap (g) of the SRR was $w = g = 10\ \mu m$ to resemble the structure of Fig. 3.4b. The red and yellow regions indicate the $ErFeO_3$ substrate (infinite thickness, dielectric permittivity $\varepsilon_{xx} = 26$, $\varepsilon_{yy} = 23.3$, $\varepsilon_{zz} = 26.2$) and Aluminum (300 nm thickness). Orange and green elements are the THz source and monitor regions. Here, nonuniform grid was used to properly cover the thin metallic layer region with more than 5 meshes. Perfect matching layer (PML) boundary condition was used to avoid reflection from the calculation boundary. The incident THz source is polarized such that $\boldsymbol{E}_{\mathrm{THz}}$ // y // gap-baring side of SRR, as indicated in (**b**)

was monitored at the center of the SRR, 10 μm below the surface of the $ErFeO_3$ substrate. The impulsive THz wave was incident from out of the substrate with its polarization aligned such that $\boldsymbol{E}_{THz}$ // gap-baring side of SRR. The resonance curves obtained from the Fourier transformation of the calculated waveforms are shown in Fig. 3.6a. As can be seen in (b), the SRR resonance frequency scales approximately inverse-proportionally to its lateral length L. The resonance frequency for the main structure used in the experiment ($L = 60\ \mu m$) is calculated to be ~0.22 THz.

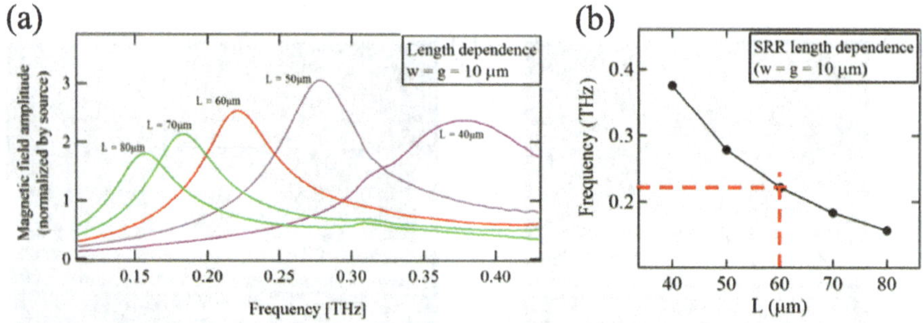

Fig. 3.6 **a** Calculated resonance enhancement curves of the SRRs with different sizes L. **b** L-dependence of the resonance frequency

3.2 Experiment Configuration and Measurement Setup

In order to excite the spin precession dynamics effectively, we have to match the resonance frequency of SRR and that of the spin system. To do this, we focus on the temperature-dependence of the FM-mode. As mentioned in Chap. 2, the $ErFeO_3$ exhibits two magnetic resonance modes, the quasi ferromagnetic (FM) and the quasi antiferromagnetic (AF) mode [Fig. 2.5]. The FM-mode can be interpreted as the precession of macroscopic magnetization around its easy axis. Because the frequency of FM-mode depends strongly on temperature, we can control the coupling strength between SRR and FM-mode by tuning its frequency around the SRR by changing the temperature.

The schematic of experimental configuration is shown in Fig. 3.7. The experiment was conducted with two mutually orthogonal incident THz polarizations. In the first experiment [Fig. 3.7a], the electric and magnetic fields of the THz radiation were parallel to the *b*- and *a*- axes of the crystal, respectively (SRR-excitation polarization). The incidence of the THz pulse with electric field polarized parallel to the gap of the SRR induces a circulating current in SRR. In this polarization, the magnetic field component of the incident THz pulse does not directly excite the spin precession. The circulating current in the SRR produces an enhanced magnetic near-field polarized along the *c*-axis, which causes the $ErFeO_3$ spins to process around the *a*-axis. Because of the precession, a magnetization component parallel to the *c*-axis appears, which can be detected via optical Faraday rotation of the visible probe. If the temperature was tuned so that the frequency of FM-mode spin precession was matched with that of SRR resonance, then the Faraday signal that corresponds to the spin precession should be enhanced at a particular temperature. In this way, we can expect to realize the excitation of spin system using magnetic nearfields of SRR.

In the second experiment, the polarization of the incident THz pulse was rotated by 90° [Fig. 3.7b; spin-excitation polarization]. Under these conditions, the electric-field component of the THz pulse does not excite the circular-current mode of SRR, but the magnetic component directly excites spin precession. By comparing the spin

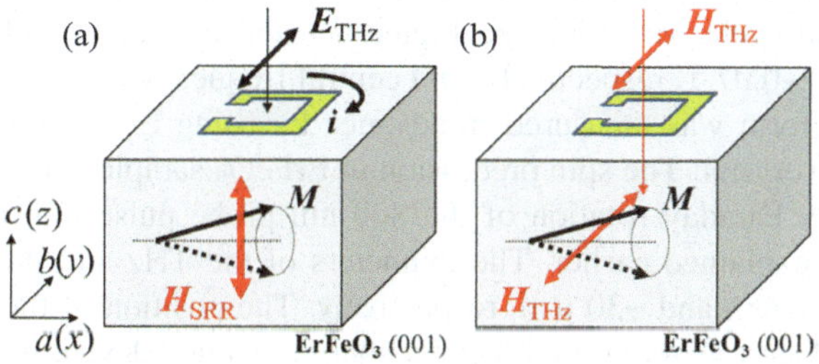

Fig. 3.7 Schematic of the experimental configuration used in the experiment. **a** The SRR-excitation polarization configuration, wherein the incident THz electric field excites the SRR which then causes spin precession mediated by its magnetic nearfield. **b** The spin-excitation polarization configuration, wherein the incident THz magnetic field directly initiates the spin precession without exciting the SRR resonance

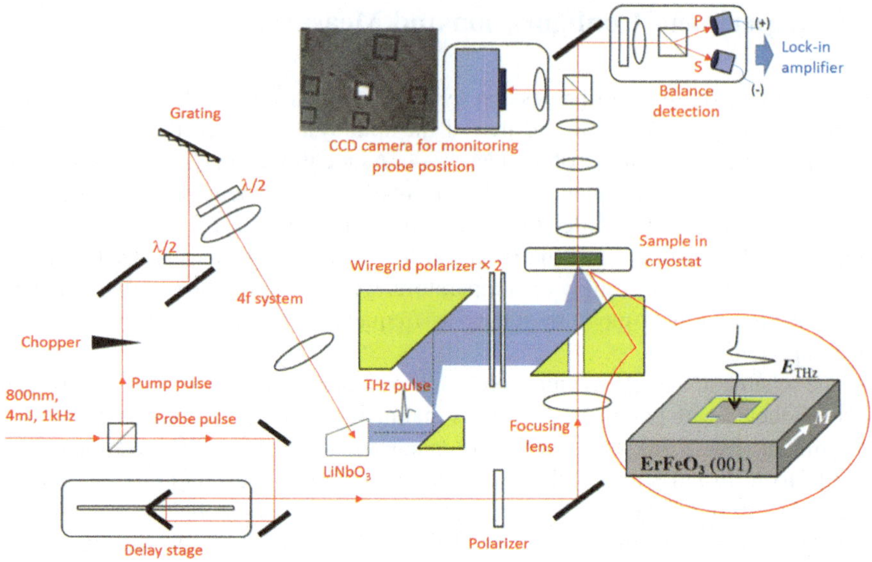

Fig. 3.8 Schematic of the experimental setup used for intense THz pump-optical Faraday probe measurement. The probe pulse can be chosen to be either focused on the sample surface for measurement of waveform (by inserting the focusing lens) or collimated to achieve microscopic observation of the magnetic domain structure (Fig. 3.3) and positioning of the probe spot (inset) by removing the focusing lens

dynamics obtained in the two configurations, we can extract the information on how the spin and SRR resonances influence each another, as will be discussed later.

The measurement was conducted with intense THz pump-optical Faraday probe setup as shown in Fig. 3.8. The THz pulse was generated by using the tilted-wave-front optical rectification technique in the $LiNbO_3$ nonlinear optical crystal as mentioned in Chap. 2. The $LiNbO_3$ was pumped by a Ti:sapphire regenerative amplifier with an energy of approximately ~3 mJ/pulse, repetition rate at 1 kHz, pulse width of ~100 fs, and central wavelength of 800 nm. The generated THz pulse had a half-cycle waveform, peak electric- and magnetic-field amplitudes of 240 kV/cm and ~0.07 T, respectively, and central frequency of ~0.5 THz (Fig. 3.9). The THz waveform was measured in advance by using EO-sampling (Fig. 2.12) with GaP (110) crystal. The spin precession in $ErFeO_3$ sample was measured in the time domain by Faraday rotation of the 800 nm probe pulse transmitted through the sample, as explained earlier. The diameters of the THz and probe spots were approximately 1 mm and ~30 μm, respectively. The position of the probe spot on the sample surface was monitored using a charge-coupled device (CCD) camera in a microscope measurement system.

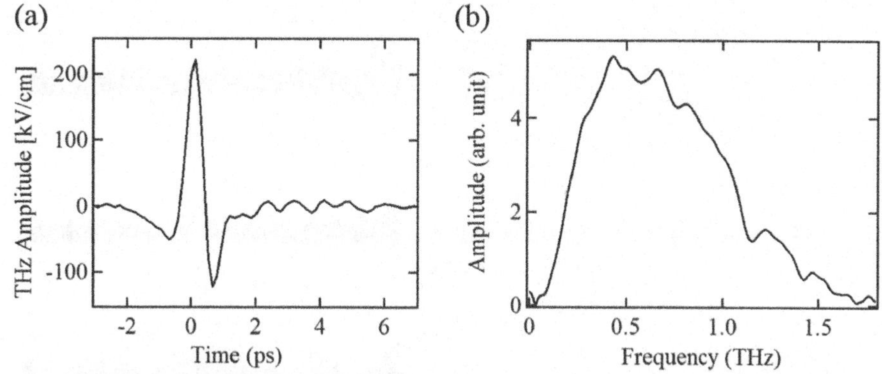

Fig. 3.9 Temporal waveform of the intense THz electric field measured inside the cryostat by using the EO sampling with GaP (110) single crystal

3.3 Result 1: Resonant Enhancement of FM Mode Precession

3.3.1 Original Spin Precession Dynamics of FM-Mode Without SRR

As the first step, we measured the original spin precession dynamics of the FM-mode in the bare substrate (without SRR structure) in the spin-excitation polarization configuration. Figure 3.10a shows the temporal waveforms measured at some typical temperatures 40, 60 and 70 K. At each temperature, the oscillation of Faraday rotation signal can be clearly seen after excited by the impulsive magnetic field of the incident THz pulse at t = 0 ps. It is also seen that this oscillation frequency lowers as the temperature raises. The peak frequency of its Fourier transform is plotted as a function of temperature in Fig. 3.11. The temperature-dependence agrees with that of the FM-mode precession previously reported [5, 6], which indicates that the observed signal actually corresponds to the THz magnetic-field induced spin precession.

In Fig. 3.10a, it can be noticed that the envelope of the FM-mode oscillation does not exhibit single exponential decay as expected from simple LLG equation but exhibits amplitude modulation (beating structure) with the period of several tens of picoseconds. Such feature is always visible in the FM-mode signal of $ErFeO_3$ in the low-temperature phase, and becomes especially clear in the lower temperatures where the lifetime of precession is much longer. As an example, the waveform at 40 K [Fig. 3.10b] exhibits a quite long lifetime that reaches over 1 ns and a clear sinusoidal envelope with ~200 ps. The origin of such modulation of the envelope function of FM-mode is unclear. As possible explanation, the emergence of some kind of standing spin waves caused by magnetostatic interaction (Walker-mode) [7], or the effect of hidden canting present in the spin sublattice is proposed [8, 9].

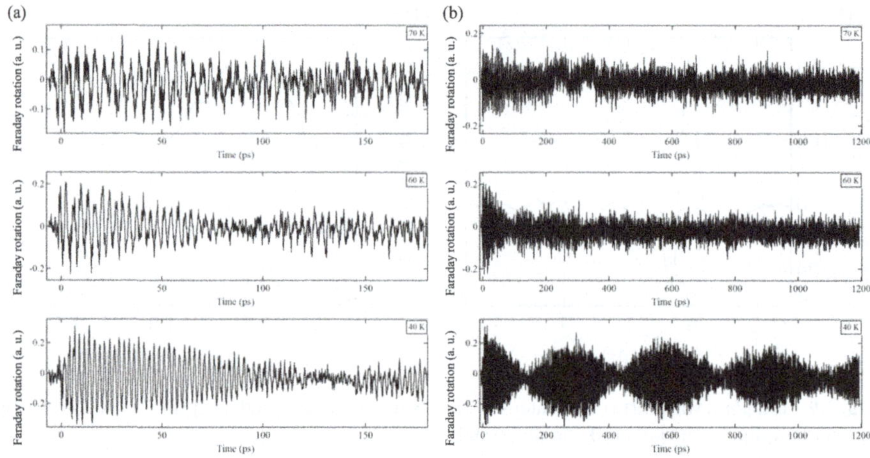

Fig. 3.10 Spin precession measured by Faraday rotation in the bare substrate (without SRR) showing oscillation in the FM-mode frequency of $ErFeO_3$. **a** Waveform shown in (**b**) expanded in the time window from −6 ps to 180 ps, and **b** overview from −6 to 1200 ps

Fig. 3.11 Temperature-dependence of the frequency of the FM-mode of $ErFeO_3$ measured by Faraday rotation at the bare substrate (without SRR). The dotted line indicates the calculated frequency of the SRR structure used in the experiment

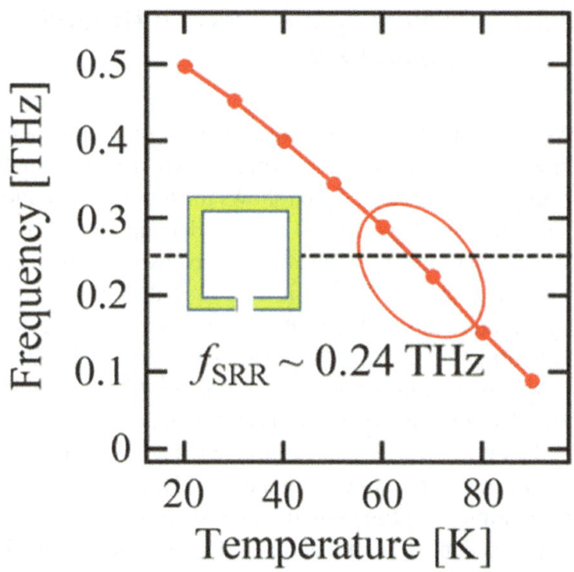

In the following, however, we conduct the measurement at the temperature region mainly above 60 K (especially near 70 K) and before 300 ps after THz excitation, wherein the beating structures are not so pronounced. In this region, most of the basic features can be well understood under the assumption that spin dynamics follows conventional single exponential-decay feature and therefore, we will not step further into the details of this beating feature.

3.3.2 Temperature-Tuning of FM-Mode Frequency Around SRR Resonance

Figure 3.12a and b show the temporal waveforms and Fourier spectra of the Faraday signal, respectively, obtained in the first experiment with SRR-excitation polarization. Here, the resonance frequency of spin precession was swept from around 0.20

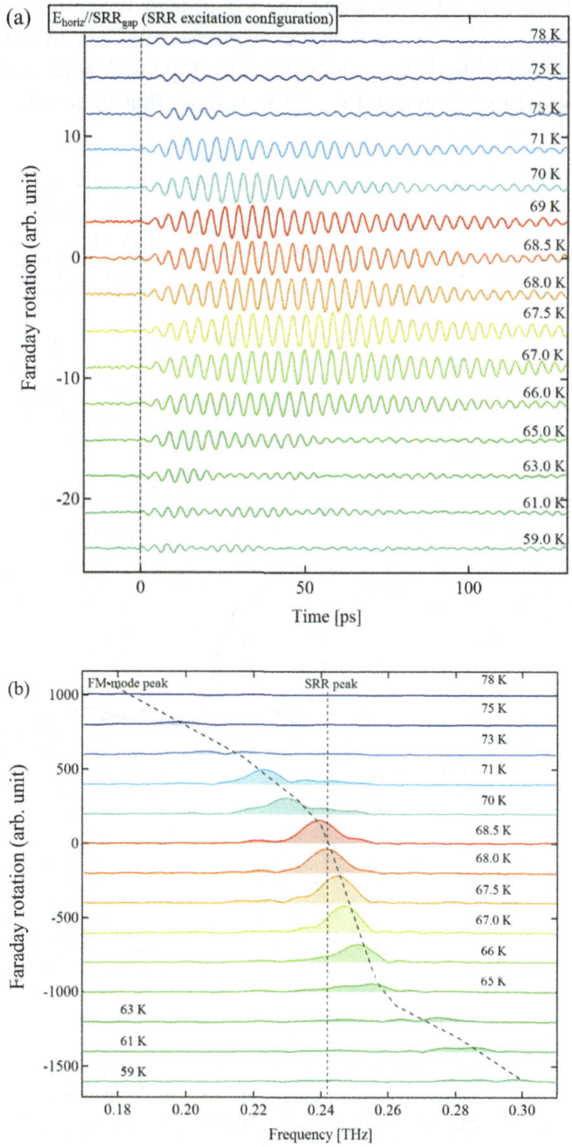

Fig. 3.12 **a** The transient Faraday waveforms measured at various temperatures in the SRR-excitation polarization **b** Fourier spectra of (**a**). The time window of [−17 ps, 133 ps] was used for transformation. The dotted line and curve indicate the resonance peaks corresponding, respectively, to the SRR and the FM-mode

to 0.29 THz by changing the temperature from 75 to 59 K. The THz pulse is incident on the sample at t = 0 ps. As can be seen, the spin precession waveform significantly changed with temperature. The most drastic change is seen in the amplitude. Clearly, the amplitude of spin precession significantly increased in the temperature region of around 68 to 69 K. The corresponding Fourier spectrum [Fig. 3.12b] revealed that the amplitude has a maximum around this temperature region. At the temperatures out of this region, we observed a smaller signal amplitude and beating of the spin oscillation in the earlier periods (0 to around 50 ps) of the time evolution waveform in (a). For example, the waveforms measured at 61 and 75 K are expanded in Fig. 3.13a. Clearly, both waveforms exhibit a beating in the early time. For example, that of 61 K shows a sinusoidal envelope with approximately 20 ps period. Its Fourier spectrum shown in Fig. 3.13b reveals that in addition to the main peak of the spin

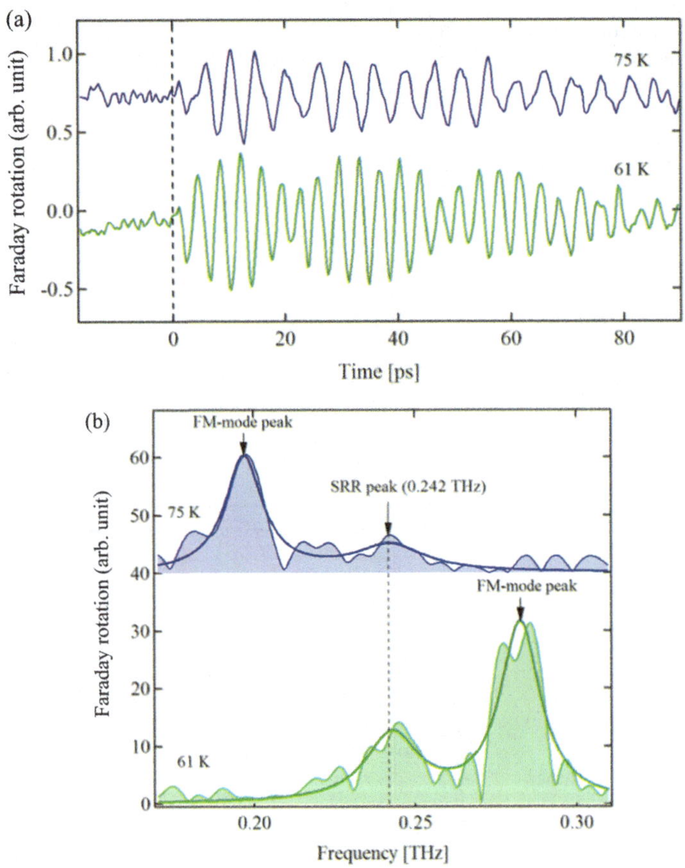

Fig. 3.13 **a** Expanded waveforms of Fig. 3.12a for the temperatures away from resonance (T = 61 and 75 K), showing beating-signature. **b** Corresponding Fourier spectra showing two peaks corresponding to the FM-mode and the SRR resonance

precession (~0.285 THz), there is a weak peak at ~0.242 THz. Such small peak also appears in the other waveforms at the same frequency, independent of temperature. This indicates that it is of different origin from the beating structure present in the bare substrate mentioned in the previous section. The frequency of this fixed peak is close to the resonance frequency of the circulating-current mode of SRR estimated in advance from the numerical calculation (FDTD). Therefore, we can ascribe this peak to be the SRR resonance frequency. The beating period of approximately 20 ps (~40 GHz) observed for the 61 K-waveform agree well with the detuning, or the frequency difference between the FM-mode spin precession peak (~0.285 THz) and SRR peak (f_0 ~ 0.242 THz).

As shown in Fig. 3.12a and b, at around 68 K where the enhancement of signals were observed, the frequency of the FM-mode crosses that of SRR resonance. This indicates clearly that the observed signal enhancement was caused by resonant excitation by a quasi-CW magnetic nearfield induced by the SRR resonance. The peak amplitude of the Fourier spectra in Fig. 3.12b are plotted as a function of detuning of the spin precession from the SRR peak in Fig. 3.15a. The full width of half maximum (FWHM) of the enhancement curve is approximately Δf ~25 GHz. From this, the quality-factor of the resonance enhancement is estimated to be $Q = (f_0/\Delta f)$ ~9.7.

On the other hand, as shown in Fig. 3.14, the spin precession amplitude excited directly by the magnetic field of incident THz pulse at a position without SRR structure was nearly an order of magnitude ($\times 1/8$ times) smaller than the maximum spectral peak amplitude obtained at 69 K in the abovementioned resonant-excitation case. This clearly indicates that the SRR magnetic nearfields actually has the ability to enhance the spin precession amplitude through resonant excitation.

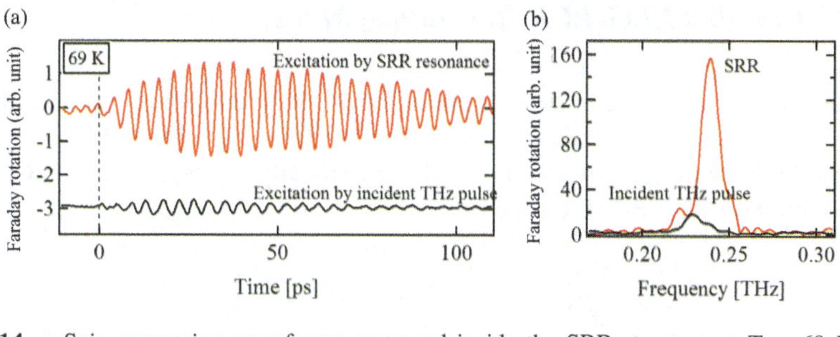

Fig. 3.14 **a** Spin precession waveforms measured inside the SRR structure at T = 69 K with the SRR-excitation polarization and a typical waveform measured outside the SRR structure (bare substrate) with the spin-excitation configuration. **b** Fourier transform of (**a**) in the range $t = [-17\,\text{ps}, 133\,\text{ps}]$

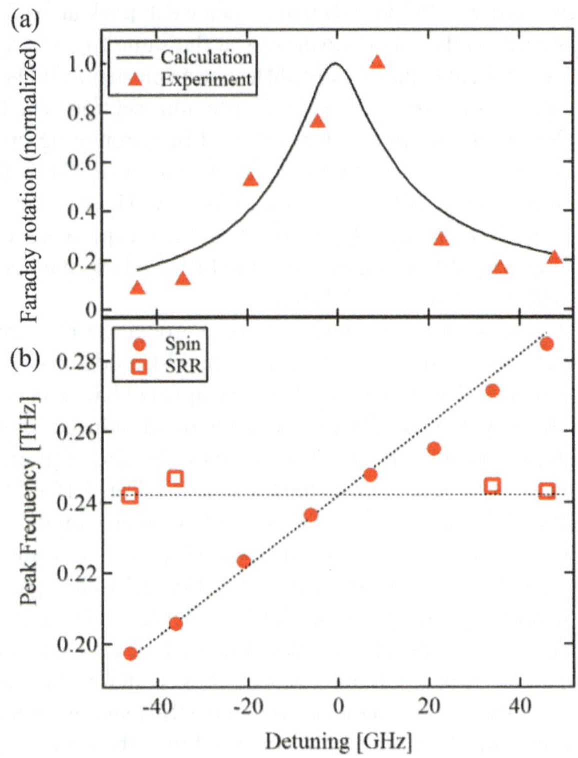

Fig. 3.15 **a** Peak amplitude of the Fourier spectra in Fig. 3.12b plotted as a function of detuning of the FM-mode from the SRR resonance frequency. The solid curve is the result of calculation by coupled LLG-LCR model. **b** Frequency of each peaks plotted as a function of detuning

3.3.3 Coupled LLG-LCR Resonance Model

In order to understand the dynamical behavior of observed spin precession in more detail, we constructed a model based on Landau-Lifshitz-Gilbert (LLG) equation to describe the spin dynamics [Eq. (3.2)], magnetically coupled to an LCR circuit, which expresses the SRR [Eq. (3.1)].

$$L\frac{d^2i(t)}{dt^2} + R\frac{di(t)}{dt} + \frac{1}{C}i(t) = \frac{dE(t)}{dt} \tag{3.1}$$

$$\frac{dM_1}{dt} = -\gamma M_1 \times H_{eff,1} + \frac{\alpha}{|M_1|}M_1 \times \frac{dM_1}{dt} \tag{3.2}$$

$$E(t) = E_{THz,y}(t) + m_1\frac{dM_{1z}(t)}{dt} \tag{3.3}$$

$$\boldsymbol{H}_{eff,1}(t) = H_0\hat{\boldsymbol{x}} + H_{THz}(t)\hat{\boldsymbol{y}} + s_1 Li(t)\hat{\boldsymbol{z}} \tag{3.4}$$

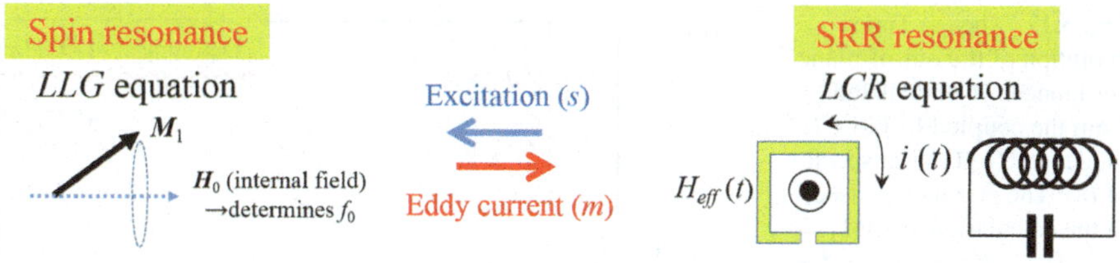

Fig. 3.16 Schematic of the bidirectional coupling between spin- (LLG equation) and SRR-resonances (LCR equation) mediated by magnetic nearfield

Schematic illustration of the model is shown in Fig. 3.16. The current $i(t)$ is driven by an external force $E(t)$ that consists of an incident THz electric-field pulse $E_{THz,y}(t)$ and also the electromotive force $m_1 \cdot dM_{1z}(t)/dt$ caused by the oscillation of the magnetic moment inside the SRR [Eq. (3.3)]. On the other hand, the magnetic moment $M(t)$ is driven by a magnetic field generated by the current $i(t)$ in the *LCR* circuit [Eq. (3.4)].

Here, the y-component of the incident magnetic field of the THz pulse $\boldsymbol{H}(t)$ is zero. The parameter H_0 expresses the static internal magnetic field (exchange and anisotropy field) of the $ErFeO_3$ crystal and determines the resonance frequency of spin precession. The parameters s_1 and m_1 determine the strength of the coupling of the field to the oscillator. The coordinates of x, y, and z correspond to the a-, b-, and c-axes of the crystal, respectively. The resistance R was estimated from the conductivity of aluminum [10, 11], while inductance L and capacitance C were determined from the resonance frequency of SRR and the Q-factor of the enhancement curve obtained in Fig. 3.15a using the following relation that is known to stand for the *LCR* circuit:

$$\omega_0 = \frac{1}{\sqrt{LC}}, \quad Q = \frac{\omega_0}{\Delta\omega} = \frac{1}{R}\sqrt{\frac{L}{C}}$$

Figure 3.17a shows the time evolution of the x-component of the magnetization M_{1z} (t) calculated for various H_0 using the presented model. The result shows a good agreement with waveforms obtained in the experiment, which indicates that the observed dynamics is well represented by the proposed framework of coupled *LLG-LCR* model. The parameters used here were determined to reproduce the waveforms observed experimentally in Fig. 3.12a. Here, the parameter H_0 was varied to tune the frequency of spin precession around the SRR resonance ($H_0 = 6.9 \times 10^6$ A/m at resonance). The corresponding Fourier spectra are shown in Fig. 3.17b. The calculated amplitude of the spin precession is shown as a function of detuning frequency in Fig. 3.15a (black solid curve). The increase of spin precession amplitude near the SRR resonance frequency is clearly reproduced.

Fig. 3.17 The (a) Time evolution of the out-of-plane component M_z calculated from the coupled LLG-LCR model and (b) Fourier spectra of (a). The parameters used in the calculation are: $|\boldsymbol{M}_1| = 1 \times 10^{-16}$ [Wb m], $m_1 = 6 \times 10^{13}$ [m^{-1}], $s_1 = -1 \times 10^{14}$ [H^{-1} m^{-1}], $\gamma = 2.2 \times 10^5$ [m s^{-1} A^{-1}], $\alpha = 0.025$, $L = 3.3 \times 10^{-11}$ [H], C = 1.33 × 10^{-14} [F] and R = 2 [Ω]

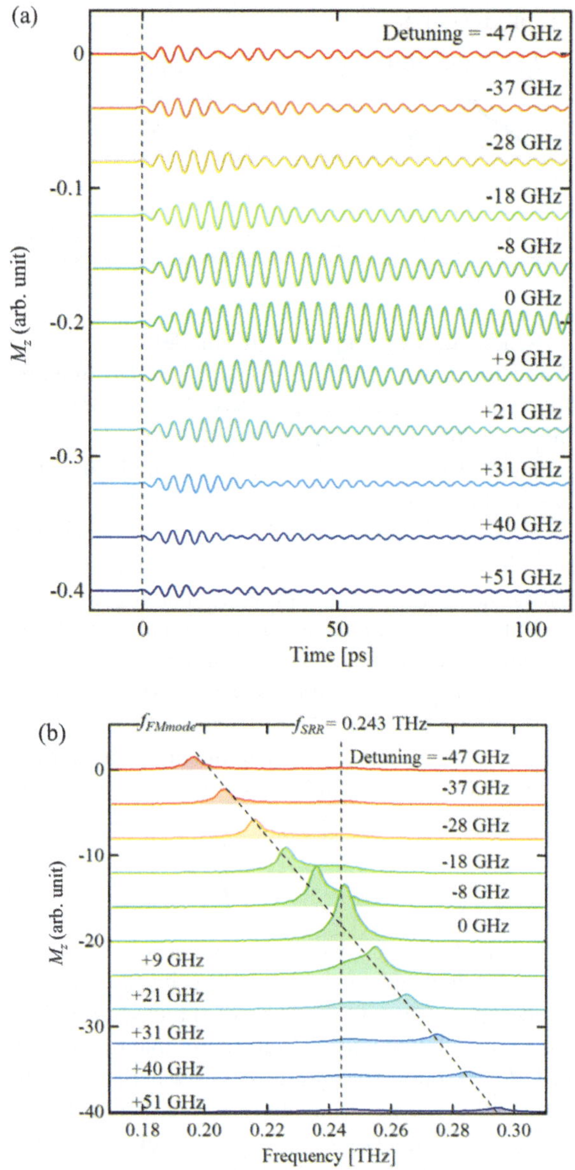

3.4 Result 2: Interactive Energy Transfer Between SRR and Spin

In the numerical model of coupled LLG-LCR equation proposed in the previous section, we introduced the term $m_1 \frac{dM_{1z}(t)}{dt}$ which describes the electromotive force induced in the SRR by oscillating magnetic moment. If we neglect this term ($m =$

0)), there will be only one-way energy transfer, that is, from the SRR to the spin precession due to excitation by magnetic nearfield H_{SRR}. In contrast, if we assume $m \neq 0$ then the introduction of such term describes the energy transfer in the backward direction, that is, the energy stored in the spin system flowing back into the SRR system through electromotive force. In this section we examine this assumption in more detail by investigating the spin dynamics at the temperature where the SRR- and spin- resonance frequencies are matched with each other, and the coupling effects between two oscillators are expected to be strongest.

For this purpose, we compared the waveform measured in the previous section for SRR-excitation polarization with that obtained for the spin-excitation configuration [Fig. 3.7b]. As mentioned earlier, with this polarization the magnetic field of the incident THz pulse directly excites the spin precession without exciting the SRR. Here, we set the frequency of the spin resonance near that of the SRR by tuning the temperature at 69 K.

The result is shown as the green curve in Fig. 3.18. For comparison, the waveform measured in the previous section with the SRR-excitation polarization (red curve) and the response of the bare substrate (without SRR, blue curve) measured with spin-excitation polarization at the same temperature are also plotted. Interestingly, clear difference can be observed in the lifetimes of the waveforms. While the waveforms with spin-excitation polarization measured inside and outside of the SRR structure exhibit oscillations that last for over 250 ps, the SRR-excited precession decays within ~160 ps. This indicates that the lifetime of spin precession is shortened when SRR structure was nearby, compared to the free-oscillation case in the bare substrate.

Another feature seen in the SRR-excitation waveform is that in the early time of oscillation, the amplitude of the waveform gradually increases from zero after excitation by incident THz pulse ($t = 0$ ps) with a finite time constant of approximately 40 ps and then continuously decreases with an ever longer time constant of

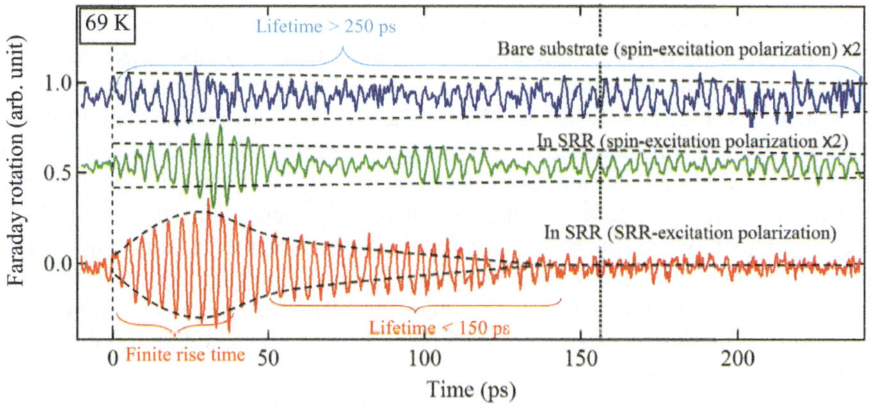

Fig. 3.18 The transient Faraday waveforms measured at 69 K with spin-excitation polarization outside of the SRR (blue), with spin-excitation polarization inside the SRR (green) and with SRR-excitation polarization inside the SRR (red). Black dotted curves are the guide to the eyes

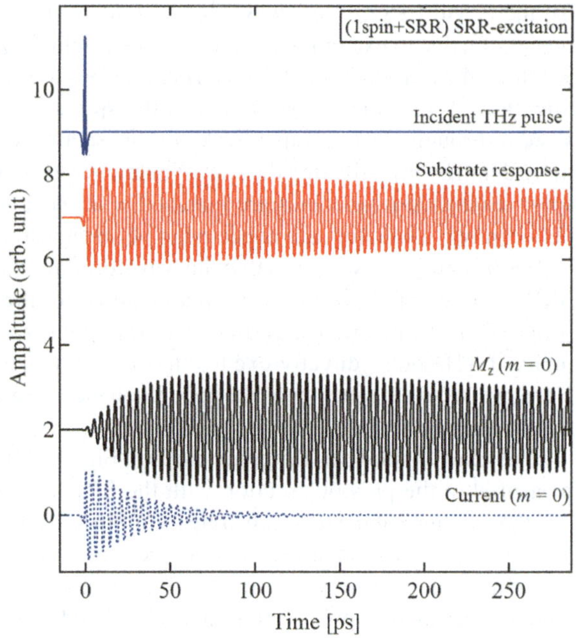

Fig. 3.19 Results of the LLG-LCR model calculation assuming the coupling term $m = 0$. Incident THz waveform for simulation (Blue solid), time evolution of M_z calculated without SRR and with spin-excitation polarization (Red), M_z excited by SRR with SRR-excitation polarization (Black), and electric current in SRR (Blue dashed)

~160 ps. In the original waveform measured in the position without SRR structure (bare-substrate) with spin-excitation polarization, such feature is not apparent.

In order to understand the abovementioned lifetime-shortening in the presence of SRR in detail, we performed a numerical calculation again using the coupled LLG-LCR model (Fig. 3.19). Here, the spin-damping parameter α was chosen to reproduce the original lifetime of spin precession measured at the bare substrate without the SRR (the blue trace of Fig. 3.18). First, we calculated the spin precession in the SRR-excitation configuration without including the electromotive force term, i.e., $m = 0$, as shown in the black trace in Fig. 3.19. Simultaneously calculated waveform of the current in the SRR structure is also indicated as the blue dotted curve.

When the THz electric field is incident on the SRR, it induces an oscillating current which decays at a time constant of ~60 ps that is determined by the Joule energy loss (R) of the LCR circuit. During the presence of such resonant current-oscillation, the spin precession amplitude M_z continuously increases. After the current oscillation has ceased completely, the spin precession amplitude is maintained at almost a constant value, then slowly decreases with a much longer decay time that corresponds to the lifetime of original spin precession without the SRR (red trace). This shows that the observed lifetime-shortening of spin precession at SRR-excitation polarization cannot be reproduced if $m = 0$.

Next, we performed the same simulation including this term, i.e., $m \neq 0$, while keeping other parameters fixed. The calculation results for the SRR-excitation and spin-excitation polarizations are shown in Fig. 3.20a and b, respectively. First we focus on the waveform in SRR-excitation polarization (a). In contrast to the previous

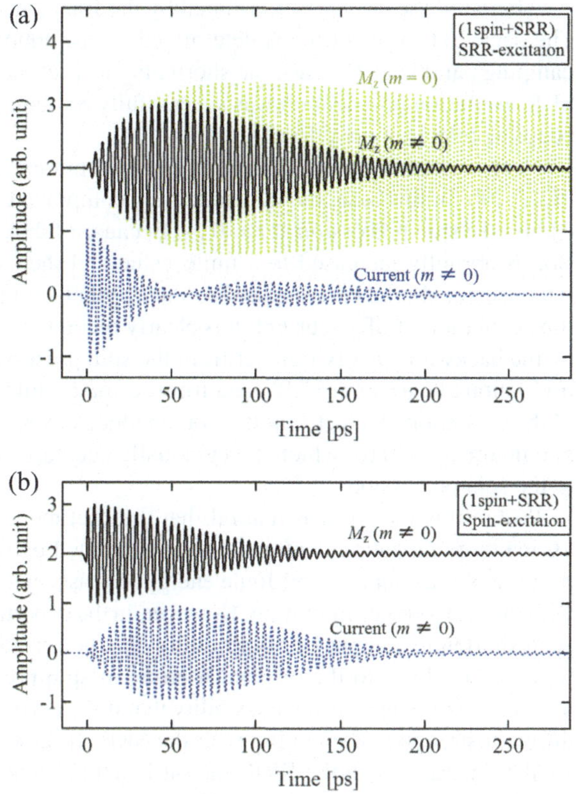

Fig. 3.20 Calculation results of the LLG-LCR model assuming the coupling term $m \neq 0$ ($m = 6 \times 10^{13}$ [m^{-1}]). In the SRR-excitation polarization (**a**), the energy once given from the SRR to the spin precession eventually flows back into the SRR. Due to this bidirectional energy transfer, the waveform calculated with $m \neq 0$ exhibits shorter lifetime compared to the case of $m = 0$. In the spin-excitation polarization (**b**), the same mechanism shortens the lifetime when $m \neq 0$

case, the spin precession waveform (black) exhibits a lifetime that is much shortened compared to the previous case of $m = 0$ (yellow). Also, another interesting feature can be seen in the temporal evolution of the current profile. After the incidence of THz pulse the current oscillation is induced in SRR and it decreases exponentially, similar to the previous case. However, the rate of this exponential decrease is slightly faster. Furthermore, a striking difference can be seen after the amplitude has reached its minimum at t ~50 ps. After this time, the amplitude of current oscillation starts to grow again. It reaches a maximum at t ~100 ps and then continuously decreases with a time constant that does not coincide with the decay rate of original spin precession (red curve in Fig. 3.19). Interestingly, the decay rate of SRR current and that of spin precession in this time region is quite similar.

The calculated features mentioned here can be summarized as follows: After the excitation of SRR by THz field, the electromagnetic energy is first given from the THz pulse to the SRR current. It is then passed to the spin precession due to resonant excitation with a time constant of ~50 ps. When all energy was passed to the spin precession, it then starts to flow back into the SRR again though the electromotive force induced by the oscillation of magnetic moments.

Afterward the energy is consumed by the Joule energy loss in the SRR structure. The decay rate at this time is determined by the coupling strength m and the Joule damping parameter R. Thus, the shortening of lifetime observed in Fig. 3.18 for the SRR-excitation polarization was successfully reproduced by taking into account the coupling (electromotive force) term m.

Next we focus on the dynamics for the spin-excitation polarization shown in Fig. 3.20b. In this case, the magnetic field component of the THz pulse impulsively excites the spin precession at $t = 0$ ps. Because of this, the amplitude of spin precession is abruptly increased to a finite value and then starts to decrease. The decay rate is again shortened compared to the original spin precession. By looking at the time evolution of SRR-current, it is clearly recognized that this shortening is caused by the backward energy transfer from the spin precession to the SRR structure. The oscillation of current in SRR starts from zero-amplitude and continuously grows due of the excitation by electromotive force induced by spin precession until it reaches a maximum at t ~50 ps, which then gradually decreases with the time constant similar to the spin precession.

Within this model, it is natural that the lifetime of spin precession in the spin-excitation polarization configuration is shortened by the presence of SRR structure because of the coupling and Joule energy loss associated with it, in a similar manner with the previous configuration. However, in the experimental waveform of Fig. 3.18, such shortening cannot be observed for this polarization. Rather, it exhibits a lifetime that is closer to the original lifetime of spin precession measured at the bare substrate. This suggests the possibility that there exist two kinds of spins which have different strengths of coupling with the SRR. If there was another spin which only weakly interacts with the SRR, then it is natural that its lifetime corresponding to the abovementioned energy transfer process from the spin to SRR system is less influenced by the SRR, giving rise to longer spin precession lifetime.

Based on this idea, we extended our coupled LLG-LCR model to include the second spin that has a weaker coupling constant with the first one. The new sets of equation then becomes:

$$L\frac{d^2i(t)}{dt^2} + R\frac{di(t)}{dt} + \frac{1}{C}i(t) = \frac{dE(t)}{dt} \tag{3.5}$$

$$\frac{dM_1}{dt} = -\gamma M_1 \times H_{eff,1} + \frac{\alpha}{|M_1|}M_1 \times \frac{dM_1}{dt} \tag{3.6}$$

$$\frac{dM_2}{dt} = -\gamma M_2 \times H_{eff,2} + \frac{\alpha}{|M_2|}M_2 \times \frac{dM_2}{dt} \tag{3.7}$$

$$E(t) = E_{THz,y}(t) + m_1\frac{dM_{1z}(t)}{dt} + m_2\frac{dM_{2z}(t)}{dt} \tag{3.8}$$

$$H_{eff,1}(t) = H_0\hat{x} + H_{THz}(t)\hat{y} + s_1 Li(t)\hat{z} \tag{3.9}$$

$$H_{eff,2}(t) = H_0\hat{x} + H_{THz}(t)\hat{y} + s_2 Li(t)\hat{z} \tag{3.10}$$

Here, Eq. (3.7) is the added LLG equation describing the motion of the second spin. Coupling parameters s_2 and m_2 were set at 1/10 of the original parameters s_1 and m_1. The lengths of the first and second spins were assumed to be the same, i.e., $|\boldsymbol{M}_1| = |\boldsymbol{M}_2|$. Other parameters were the same as in the previous calculations.

The results calculated using the coupled 2-spin and LLG model for the SRR-excitation and spin-excitation polarizations are shown in Fig. 3.21**a** and **b**, respectively. In the SRR-excitation polarization (a), because of the smaller coupling constant of the second spin, its amplitude of precession is much smaller than that of the first spin. The dynamics of the total magnetization $M_{1z} + M_{2z}$ is therefore dominated by the first spin, which is strongly coupled with the SRR.

However, in the spin-excitation polarization (b), the dynamics of total magnetization is quite different from the previous case and the oscillation lifetime is closer to the original spin precession in the bare substrate. In this case, the elongated lifetime compared to the aforementioned result of 1-spin model is ascribed to the dynamics of the newly introduced second spin (yellow curve in (b)). In this polarization, the incident THz pulse instantaneously excites the first and the second spin with an equal strength. Therefore, they exhibit similar amplitude at the first stage of oscillation around $t = 0$. However, while the oscillation amplitude of the first spin with stronger coupling with SRR decreases due to the energy transfer to the SRR, that of the second spin is almost unaffected by such effect. Due to the weaker coupling constant, the energy once given to the second spin by the incident THz magnetic pulse is stored in it for a long time without being consumed by the Joule energy loss in the SRR system. Thus the second spin dominates the lifetime in the longer time region for this polarization configuration.

We can explain the existence of such strongly- and weakly-coupled spins by considering the spatial distribution of the magnetic nearfield around the SRR structure along the direction of sample thickness. The magnetic field created by the SRR structure is localized in its close proximity whose specific spatial dimension is determined mainly by the size of the ring and the width of the SRR stripe. As mentioned previously in the experiment setup section, in this experiment the SRR with the size of $60 \times 60\,\mu$m, gap of $10\,\mu$m and stripe width of $10\,\mu$m was used. In such structure, the magnetic field is expected to extend in the region of around 20–30 μm from the SRR fabricated on the surface of $ErFeO_3$ crystal. Because the thickness of the crystal was 130 μm, there exist a vast region below the SRR where its magnetic nearfield is not influential. In such region, the coupling of the magnetic nearfield with the spin system is expected to be weaker than at the surface.

Finally, it is worth mentioning that the abovementioned energy transfer from the electromagnetic radiations from excited states of the electronic systems of matter to the external cavity-mode in the nearfield region has close resemblance to the effect well known as Purcell effect [130, 131], wherein a fluorescent particle placed in the vicinity of a metallic mirror or an optical cavity exhibits a much shortened lifetime of luminescence compared to the free-space environment, due to the increase of the

Fig. 3.21 Calculation results of the extended LLG-LCR model assuming 2 spins and $m \neq 0$. In the SRR-excitation polarization (**a**), the dynamics of the total magnetization is dominated by the strongly coupled spin M_1. In contrast, in the spin-excitation polarization (**b**), the lifetime is dominated by the weakly-coupled spin

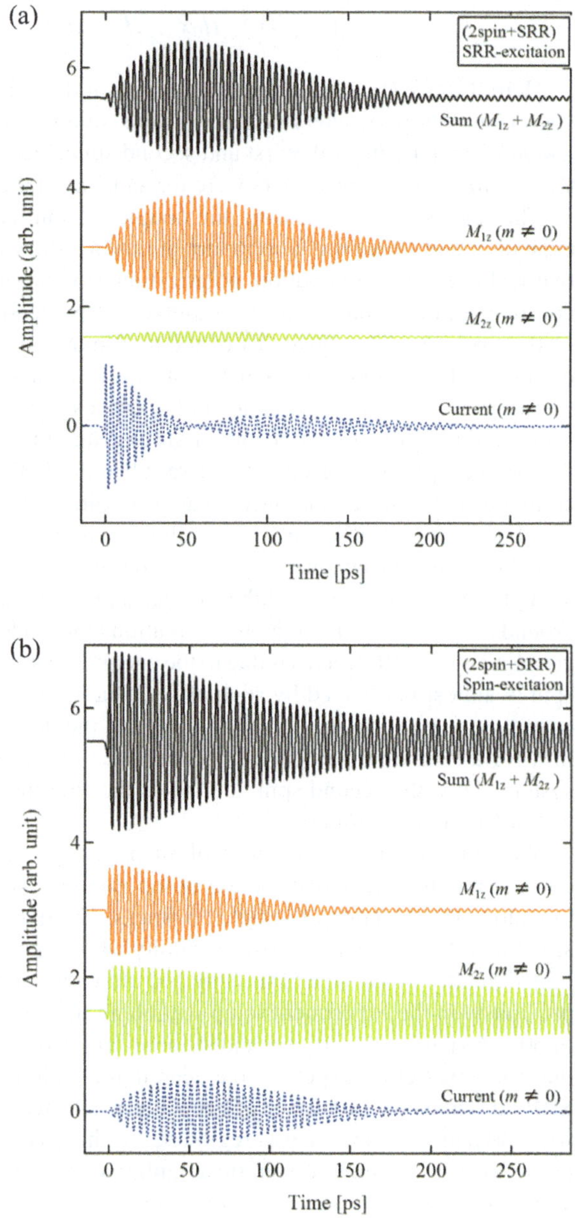

radiative decay rate of the excited state by the evanescent coupling to the external cavity mode or surface plasmon mode of the metal surface. We can view our case of spin-precession under SRR as a similar phenomenon, in which the energy transfer from the spin to the SRR was mediated by the evanescent THz magnetic nearfield, and caused the loss of excited-state energy of the spin system.

3.5 Chapter Summary

In conclusion, in this chapter we studied the spin-precession dynamics in orthoferrite $ErFeO_3$ magnetically coupled to the resonance mode of SRR using a THz pump-optical probe setup. We observed a spin precession amplitude that is enhanced by nearly an order of magnitude at maximum in the resonance excitation temperature region. Through polarization measurement and numerical calculation using a coupled LLG-LCR equation, it was suggested that bidirectional energy flow between spin precession and SRR resonance mediated by the magnetic nearfield is essential in understanding the dynamical behavior of the spin precession. Such coupling behavior with the SRR cavity observed in the spin precession dynamics has not been reported in the past, and suggests the potential applicability of the electromagnetic resonances in metamaterial structure to realize sophisticated control of magnetic excitations of electronic system in the THz region.

References

1. Wood DL, Remeika JP, Kolb ED (1970) Optical spectra of rare-earth orthoferrites. J Appl Phys 41:5315
2. Bazaliy YB, Tsymbal LT, Kakazei GN, Wigen PE (2004) The role of erbium magnetization anisotropy during the magnetic reorientation transition in $ErFeO_3$. J Appl Phys 95:6622
3. Belyaeva AI, Baranova KV (2009) Role of Er3+ ion in the formation of the $ErFeO_3$ magnetic properties in the region of spin-reorientation phase transition. Bull Russ Acad Sci Phys 73:1056
4. Yee KS (1966) Numerical solution of initial boundary value problems involving Maxwell's equations in isotropic media. IEEE Trans Antennas Propag 14:302
5. Bertaut EF, Mareschal J (1967) Etude de la structure magnetique des chromites d'erbium et de neodyme par diffraction neutronique. Solid Stat Commun 5:93
6. Yamaguchi K, Kurihara T, Minami Y, Nakajima M, Suemoto T (2013) Terahertz time-domain observation of spin reorientation in Orthoferrite $ErFeO_3$ through magnetic free induction decay. Phys Rev Lett 110:137204
7. Walker LR (1957) Magnetostatic modes in ferromagnetic resonance. Phys Rev 105:2
8. Yamaguchi K (2013) Ultrafast spin dynamics induced by pulsed terahertz wave in canted antiferromagnet. Ph.D. Thesis, The University of Tokyo

9. Yamaguchi K, Kurihara T, Watanabe H, Nakajima M, Kato T, Suemoto T (2014) Beating of terahertz pulse induced spin precession in $ErFeO_3$. In: Proceedings of 19th International Conference on Ultrafast Phenomena, pp 08
10. Ordal MA, Bell RJ, Alexander LLLRW Jr, Querry MR (1985) Optical properties of fourteen metals in the infrared and far infrared: Al. Appl Opt 24:24
11. Laman N, Grischkowsky D (2008) Terahertz conductivity of thin metal films. Appl Phys Lett 93:051105

Chapter 4
Control of Macroscopic Magnetic Order Dynamics Using SRR-Enhanced THz Magnetic Fields

Abstract In this chapter, we combine the magnetic-field enhancement effect by SRR introduced in previous chapter with the optical excitation of spin systems with ultrafast laser. We focus on the spin-reorientation phase transition which occurs in $ErFeO_3$. Because the spin system is quite unstable near the phase transition temperature, it is expected that the weak fluctuation of magnetization caused by THz magnetic field in the initial stage of ultrafast magnetic phase transition leads to a gigantic change in the final state. As we show in the following, with such a concept we succeeded in generating a macroscopically aligned magnetization state by coherently selecting the path of phase transition process through spin precession-induced symmetry breaking.

Keywords Spin reorientation · Symmetry breaking · Macroscopic magnetization · Ultrafast heating

4.1 Background: Controlling the Path of Phase Transition by the Coherent Spin Precession

Before describing the principle of our experiment, we will first start by briefly reviewing the previous research which focused on the optical control of SRPT process. As mentioned in Chap. 2, the magnetic structure of the orthoferrite is determined by a subtle balance between magnetic anisotropy fields of each crystallographic (*a*-, *b*- and *c*-) axes. Due to the difference in the temperature-dependence of anisotropy fields in each directions, many orthoferrites exhibit the rotation of the easy-axis of antiferromagnetic order at a particular temperature. This phenomenon is known as spin-reorientation phase transition (SRPT).

It is known that the SRPT can be driven by change of the ambient parameters such as temperature, magnetic fields or electric fields. Also, the ultrafast heating caused by irradiation of femtosecond laser transient can induce the rotation of magnetization as well. For example, in ref [1], de Jong et al. showed that by applying the fs laser pulses on orthoferrites (here, $(SmPr)FeO_3$) it is possible to induce the magnetization rotation within a timescale of picoseconds through ultrafast heating. Furthermore, they also showed that the direction of magnetization in the final state after SRPT can be switched by selecting the helicity of the incident fs pulse (Fig. 4.1).

T. Kurihara, *Observation and Control of Magnetic Order Dynamics by Terahertz Magnetic Nearfield*, Springer Theses, https://doi.org/10.1007/978-981-16-8793-8_4

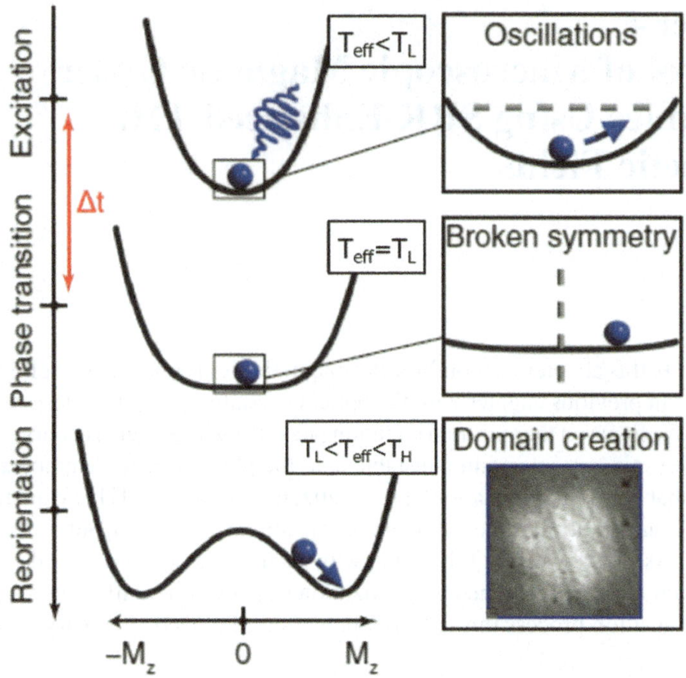

Fig. 4.1 Schematic of the mechanism of ultrafast heat-assisted macroscopic domain formation due to coherent spin precession induced by inverse Faraday effect in (SmPr)FeO_3 [1]. Figure reprinted with permission from [1], © 2012 AIP

This phenomena is explained in the following manner: when a circularly polarized fs pulse is incident on the spin system, it passes the angular momentum to the spin system (inverse-Faraday effect). Because of this, the magnetization starts to coherently process around the original easy-axis. At the same time, the thermal energy given from the incident pulse gradually heats up the spin system, changing the shape of free energy curve with time. The succeeding SRPT favors the direction at which the spin system is at the moment the sample has reached the threshold temperature T_L, because the magnetic symmetry is instantaneously broken at this time due to the tilting of magnetization caused by coherent spin precession. This results in the formation of macroscopically aligned magnetization towards either up- and down direction after the completion of SRPT process.

4.2 Motivation

Important aspect of the above process may be summarized in two fold effects: (i) instantaneous heating of the spin system by fs pulse to cause SRPT, and (ii) excitation of coherent spin precession that is used to select the path of the phase transition. As they showed clearly, such selection of the phase transition path can be realized by using a coherent spin precession with relatively small amplitude. Therefore, we can expect that the THz magnetic field can similarly select the path of SRPT through excitation of the coherent spin precession, if properly assisted by the optical heating pulse.

In the previous experiment shown above, the effects of (i) heating and (ii) spin precession were caused simultaneously by using a single circularly-polarized pump pulse. Instead, we divide this into two separate pump pulses of the THz and fs laser, so that each pulse separately bares the role of (i) and (ii), respectively.

In this way, we can expect to create a macroscopically aligned magnetic domain by coherently controlling the path of SRPT by using THz magnetic fields. Also, owing to the fact that the THz magnetic pulse can excite only the spin precession without causing unnecessary heating on the electronic systems, it is expected that we can examine the hypothesis proposed in the previous study that the direction of SRPT is determined by the transient phase of the spin precession in a much more clear-cut manner.

The schematic of our concept is illustrated in Fig. 4.2a and b. (1) At first, the $ErFeO_3$ sample is set at a temperature immediately below that of SRPT region. Here, we irradiate the sample with THz magnetic field, which causes (2) coherent spin precession in the sample. (3) Then, with a proper delay time *dt* we apply another pump pulse of fs optical heating pulse. This pulse (4) causes SRPT, which rotates the magnetization towards either up- or down directions, depending on the arrival time of the heating pulse with respect to the THz pump applied at the initial stage.

4.3 Experiment Setup/Sample Properties

4.3.1 Design of the SRR Structure

In order to obtain sufficient precession amplitude in the first stage, we used the intense THz pulse train that exhibits peak electric field of ~300 kV/cm in the cryostat, generated by optical rectification in $LiNbO_3$ crystals. Also, to further increase the magnetic field strength, we used the SRR structure similar to the previous Chap. 3. As will be discussed in the following, because of the fact that the spin precession frequency of FM mode becomes softened due to SRPT, near the SRPT temperatures the incident THz pulse exhibits only a weak spectral component at the frequency of FM mode (below 0.1 THz). We overcome this problem by using the resonant nearfield enhancement by SRR.

In the above schematic of SRPT control, the initial temperature of the sample needs to be set at immediately below the SRPT temperature region, which is between $T_L = 86 \pm 1$ K and $T_H = 97 \pm 1$ K for our $ErFeO_3$ sample. Figure 4.3 shows the

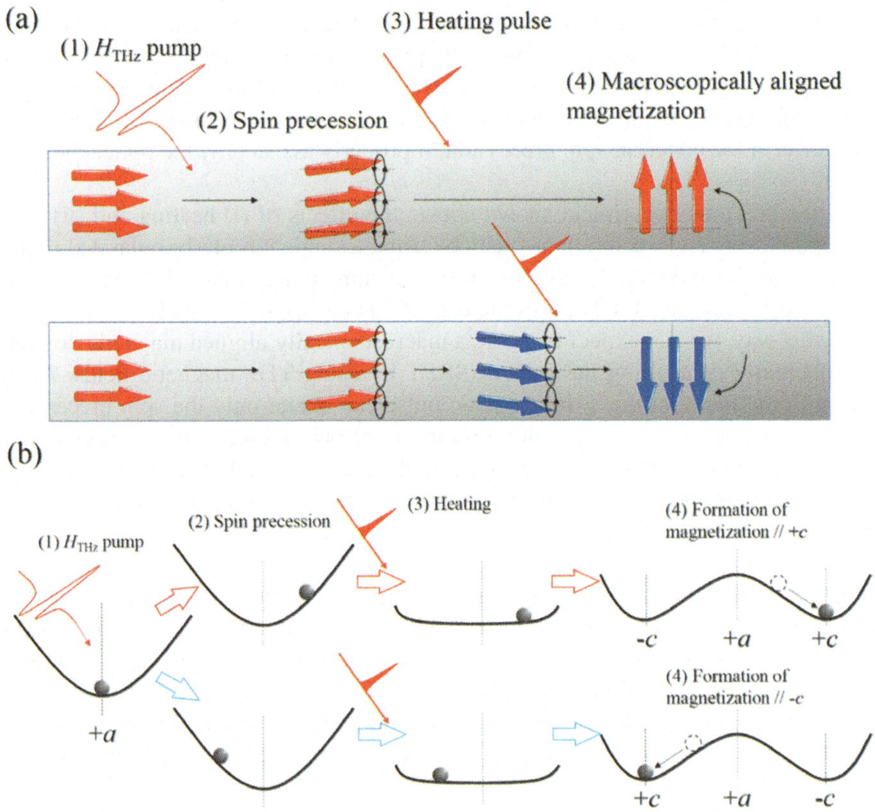

Fig. 4.2 **a** Schematic of the control of phase transition path of SRPT process. The coherent spin precession induced by THz magnetic fields (1–2) and subsequently incident optical heating pulse (3) results in the formation of macroscopically aligned magnetization domain (4) in the final state after the completion of SRPT. The direction of magnetization in the final state can be coherently selected by tuning the delay time of the incident heating pulse with respect to the THz pulse. **b** Free energy description of the process shown in (**a**)

temperature-dependence of the FM mode frequency of $ErFeO_3$ obtained by THz-TDS transmission measurement. It can be seen that near the SRPT temperature T_L ~ 86 K, the FM frequency f_{FM} rapidly softens (Owing to the accuracy of fitting, estimated value of the T_L contains an error bar of around ± 1 K). In order to realize complete rotation under finite energy of heating pulse, we have to maintain the sample temperature as close to T_L as possible. If the allowed temperature difference from T_L is set to below 5 K, for example, then the resonance frequency of the SRR structure f_{SRR} has to fit within the region from 0.12 to 0 THz.

We designed the SRR structure that exhibits resonance in such frequency region with FDTD simulation. The geometry and conditions of calculation are indicated in Fig. 4.4.

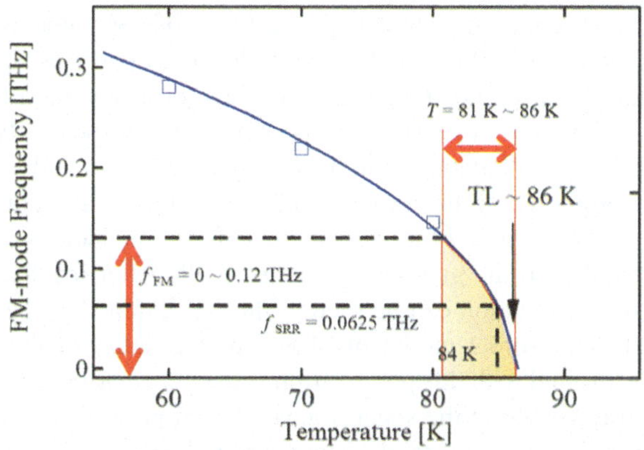

Fig. 4.3 Temperature-dependence of the frequency of the FM-mode spin precession in $ErFeO_3$, obtained by THz-TDS transmission measurement. The theoretical curve used for the fitting is $f_{FM} = A(T_L - T)^{1/2}$, as derived previously in Eq. 2.11. The lower threshold temperature is estimated to be T_L ~86 K ± 1 K

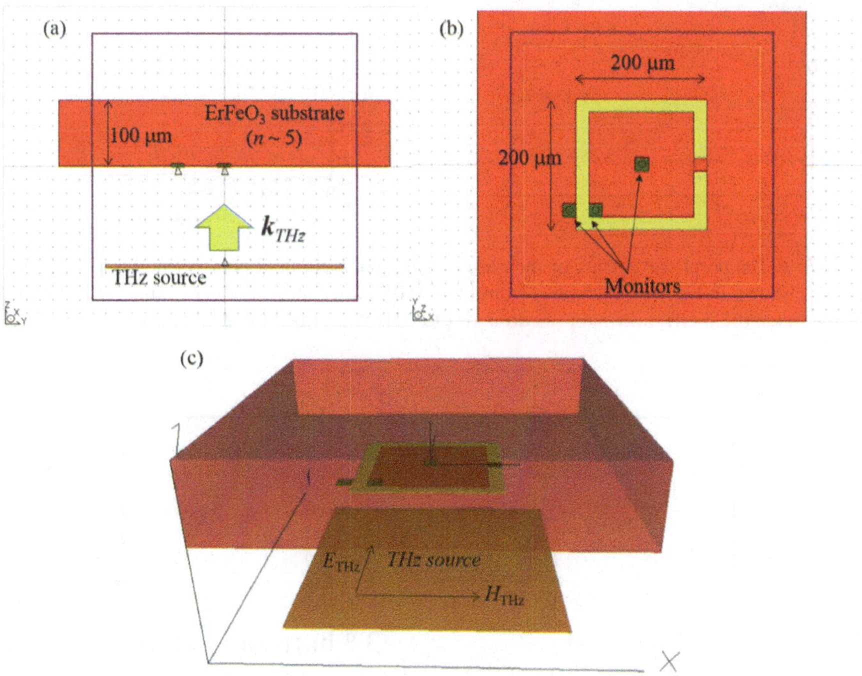

Fig. 4.4 Geometry of the FDTD calculation viewed from **a** x-, **b** z- and **c** obliquely upward direction. Essentially the same calculation configurations as in Fig. 3.5 have been used. Here, instead of Al, Au (300 nm thickness) was used as the material for SRR structure. Thickness of the $ErFeO_3$ substrate was assumed to be 100 μm to meet experimental conditions

As a result, the SRR made of gold with the thickness of ~ 300 nm, lateral dimension of 200 × 200 μm, gap of 20 μm and stripe width of 20 μm was shown to exhibit a resonance frequency f_{SRR} ~0.0625 THz. The waveforms and their Fourier spectra obtained in the calculations are shown in Fig. 4.5a and b, respectively. The incident THz pulse used in the calculation is indicated by blue curve, and the out-of-plane magnetic fields monitored at the corner (red), center (black) and outside (green) of the SRR structure (see Fig. 4.7) are also indicated. The incident THz electric field was polarized parallel to the gap-baring side of SRR. After irradiation by THz, a complex pattern of oscillation can be seen in the H_z obtained at the corner, due to the excitation of higher-order modes of SRR. The frequency of the second lowest mode that can be excited with this polarization is the third-harmonics mode depicted in Fig. 4.6b, and is visible in the spectrum as a broad peak centered at ~0.19 THz [Fig. 4.5b]. As time evolves, this mode disappears and only the sinusoidal oscillation with frequency of ~60 GHz remains, which can be assigned to the fundamental mode.

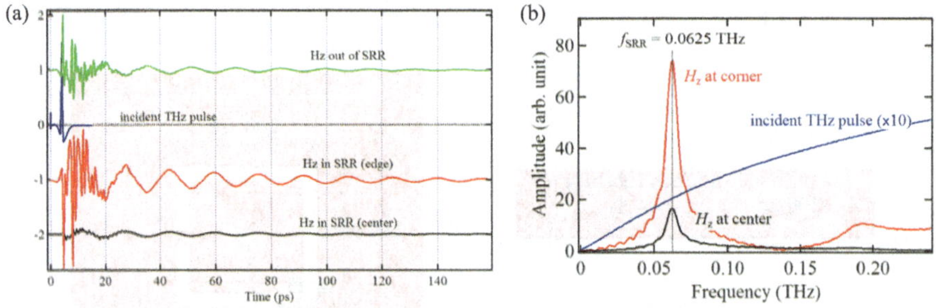

Fig. 4.5 **a** Temporal waveforms of the out-of-plane magnetic field component H_z around SRR structure shown in Fig. 4.4, calculated from FDTD. **b** Fourier spectra of (**a**) around the oscillating current mode (62.5 GHz). Broad peak around 0.19 THz is ascribed to a higher resonance mode

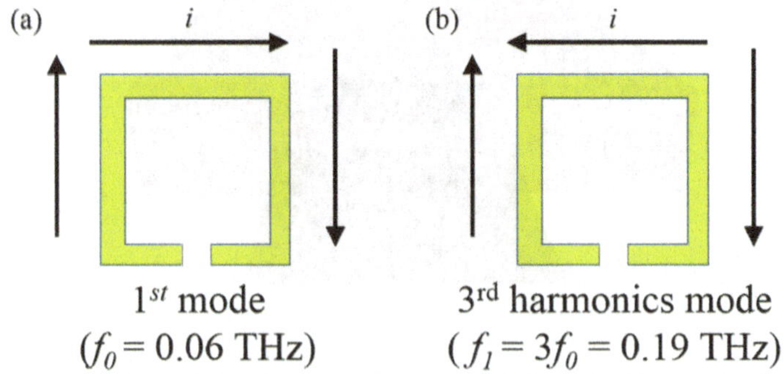

Fig. 4.6 Schematic of the resonance mode of SRR. **a** Fundamental (circular-current) mode and **b** 3rd harmonics mode

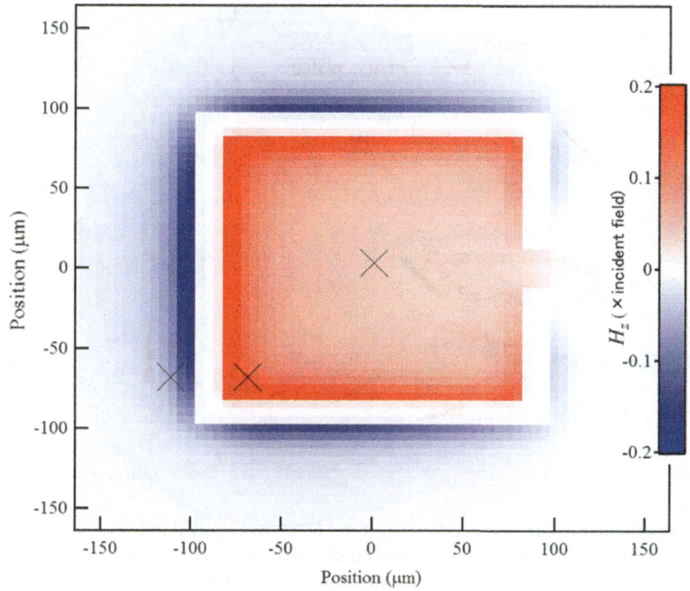

Fig. 4.7 The spatial distribution of the out-of-plane magnetic field H_z calculated from FDTD simulation. The red and blue colors indicate the direction of H_z

The spatial distribution of H_z obtained at the moment where only the fundamental mode remains is depicted in Fig. 4.7. It can be seen that at the corner of the SRR the amplitude is larger than at the center, because the distance from the metallic stripes are closer. Also, it can be seen that the direction of H_z is inverted over the SRR stripe. The Fourier spectrum Fig. 4.5b shows that at the center of the SRR structure, the spectrum amplitude of H_z at 0.065 THz is enhanced compared to the incident THz pulse by around 8 times. This factor at the corner of the SRR is even larger, reaching almost ~40 times. Therefore, it is predicted that by using the enhancement in SRR structure, the weak spectral amplitude of the incident THz waves in these frequency regions can be compensated.

4.3.2 Experimental Setup and Sample

The experimental configuration is shown in Fig. 4.8a. The measurement setup (Fig. 4.9) is similar to that used in the previous Chap. 3. However, in addition to the intense THz pump and optical Faraday probe, here we constructed another optical path of the heating pulse. The spot size of the THz pump was ~500 μm in FWHM and thus covers the whole dimension of the single SRR structure. The power of the heating pulse is ~1 μJ/pulse, and is focused down to the diameter of approximately 100 μm. As shown in Fig. 4.8b, the arrival time of the THz pulse peak was defined

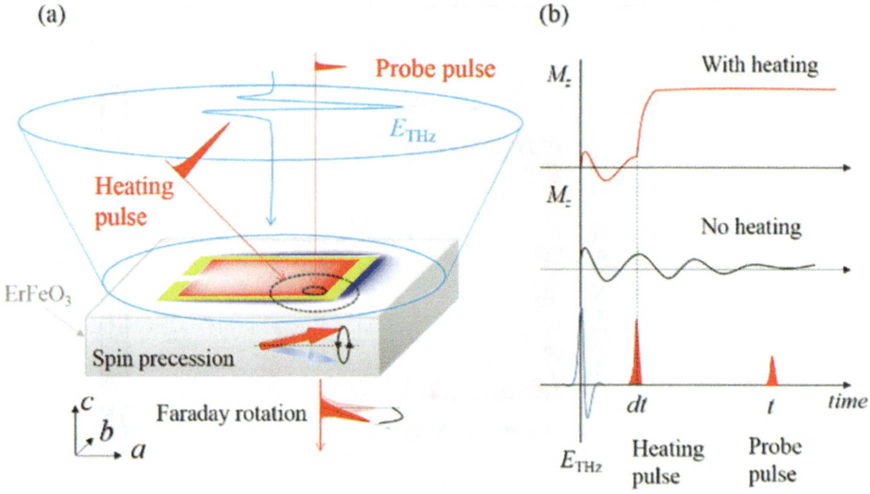

Fig. 4.8 **a** Schematic of the experiment setup for THz pump-optical pump- optical Faraday probe measurement. The inset image shows the spatial distribution of H_z calculated by FDTD (same as in Fig. 4.7). The probe spot (diameter ~15 μm) was focused at mainly near the corner of the SRR structure, where the magnetic nearfield is expected to be strongest. **b** Schematic of the time sequence of the incident pulses (bottom), THz-induced out-of-plane magnetization M_z without heating pulse (middle) and M_z after irradiation by heating pulse (top). The timing delay dt between THz and optical pulses are defined such that dt = 0 when two pulses are simultaneously incident and dt > 0 when optical pump arrives later than the THz pulse

to be $t = 0$. The arrival time of heating pulse (dt, measured from the arrival time of THz peak) was delayed by a mechanical stage, and defined such that $dt > 0$ when heating pulse arrives at the sample later than the THz. The Faraday signal induced by THz pulse was measured by a balance detector which is connected to a lock-in amplifier.

The sample used in the study is a c-cut plate of $ErFeO_3$ single crystal grown by the floating-zone method and crystal axes direction were determined again by X-ray Laue diffraction as in Chap. 3. The SRR structure was fabricated by photolithography technique, using a maskless-lithography machine in the cleanroom facility in Riken, in collaboration with the Quantum Nano-Scale Magnetism Team (Prof. Otani group). It is made of a 300 nm-thick gold film with a ~ 10 nm-thick titanium film as buffer layer, both deposited by electron-beam evaporation. Increase of metal thickness (200 nm → 300 nm) and usage of the lower resistance materials as the main metal film (Al → Au) and the buffer layer (Ni:Cr → Ti) compared to the previous Chap. 3 enabled us to further increase the Q-factor of the SRR resonance. We fabricated multiple structures on the same sample as shown in Fig. 4.10a, and used the best SRR where the largest response was observed. Main SRR used in the following experiment is indicated in Fig. 4.10b.

We should note that no external magnetic field was applied to the sample during the experiment, in order to avoid unexpected influences of such fields. Therefore,

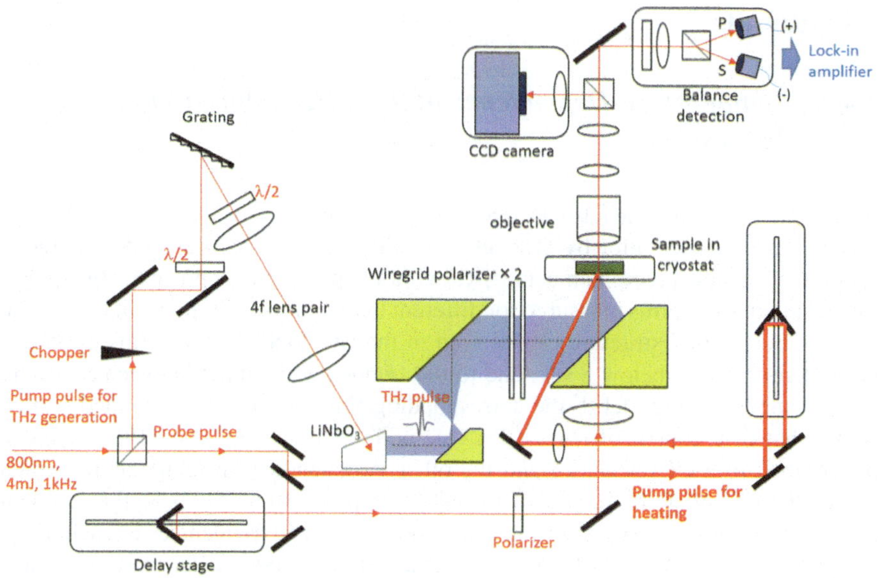

Fig. 4.9 Optical system used in the study. In addition to the intense THz pulse generation setup using $LiNbO_3$ crystal and optical Faraday probe that are used in the previous chapter, here the path of optical pump was newly added

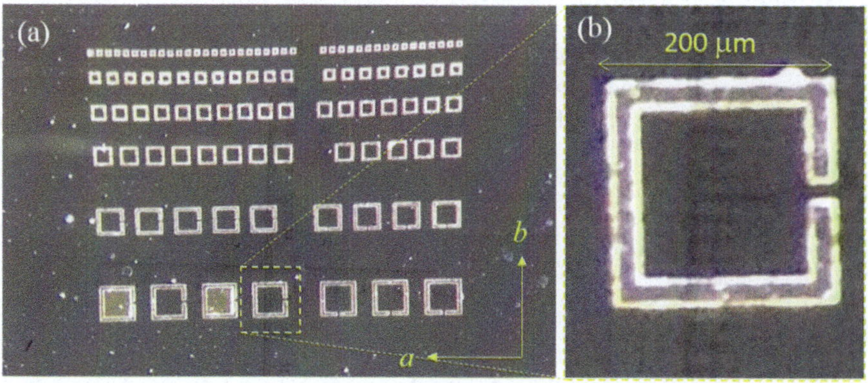

Fig. 4.10 **a** Optical micrograph of the fabricated SRR patterns, **b** Main SRR element used in this chapter with size $L = 200$ μm, width = 20 μm and gap = 20 μm

the initial state of magnetization in the sample is expected to contain $\boldsymbol{M} \,//\, +a$ and $-a$-domains simultaneously. However, as long as we excite the spin system with the $\pm c$-polarized magnetic field of SRR and measure the resulted magnetization along the same direction, the spin precessions induced in the oppositely magnetized domains are always in-phase. Therefore, the coexistence of the opposite domains does not cause any negative effects on the following experiment.

4.4 Results

4.4.1 Temperature-Dependence of the THz-Induced Spin Precession

At first, we checked if the fabricated SRR was actually working fine to excite the spin precession upon irradiation by THz pulse. For this purpose, we measured the Faraday rotation of the spin precession in the SRR without applying the optical heating pulse. The resulted waveforms measured for different temperatures are shown in Fig. 4.11a. Clearly, the beating structure can be seen in the envelope of waveforms measured at the temperatures below 83 K, due to the frequency detuning between SRR and the FM-mode. In Fig. 4.11b the corresponding Fourier spectra are plotted, which shows that they consist of two peaks. One is fixed at around 0.06 THz which can be ascribed to SRR resonance and the other is dependent on temperature, which agrees the FM-mode frequency of $ErFeO_3$ spin precession. At 84 K, the resonant enhancement of the spin precession can be observed. The frequencies of each peaks are plotted in Fig. 4.11c, which shows clearly that the two peaks cross with each other at around 84 K. These observed features well agree with the result obtained in the previous chapter. Thus, it was shown clearly that the fabricated SRR is actually capable of driving the spin resonance at a designed temperature of immediately below the SRPT. Also, from the temperature-dependence of the FM-mode frequency, the lower-temperature boundary of SRPT T_L for the present sample was estimated to be T_L ~85 K ± 1 K.

In should be mentioned that in contrast to the previous chapter, the amplitude of the spin precession remains relatively large, even at the off-resonant temperatures away from 84 K (e.g., at 78 K the FM-mode frequency is nearly an octave higher

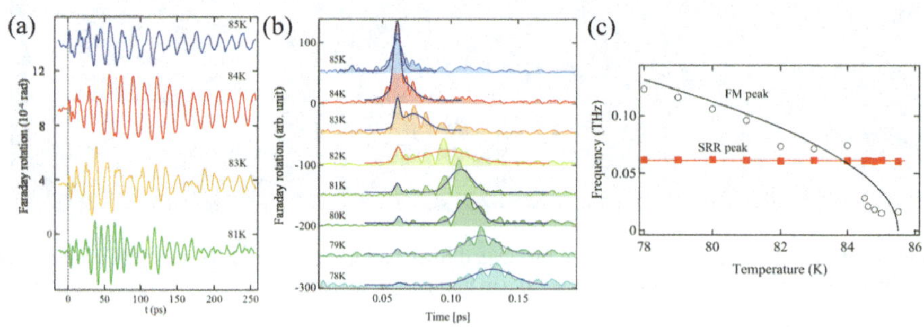

Fig. 4.11 **a** Temporal waveforms of the SRR-induced spin precession measured without heating pulse, **b** Fourier spectra of (**a**), along with fitting by two Lorentzian functions for each curves. Time window = [−13 ps, 187 ps] was used. **c** Temperature-dependence of the frequencies corresponding to the FM-mode and SRR peaks in (**b**). From fitting by the same function as in Fig. 4.3, T_L for the present case was estimated to be T_L ~ 85 K ± 1 K

than the SRR driving frequency, yet its spectral amplitude is comparable to the near-resonant temperatures such as 83 K). This is not caused by the low quality factor of the fabricated SRR structure, because the linewidth of SRR peak at 0.06 THz is sharp enough. It is suggested, therefore, that there exists some impulsive-excitation component along with the resonant excitation field due to main SRR mode.

This was not caused by the impulsive THz magnetic field contained in the original incident THz pulse, because in this SRR-excitation polarization it cannot directly excite the FM-mode. We suspect that the high order resonance modes of the SRR is responsible for this. As we have seen in Fig. 4.5b, the 3rd harmonics mode of SRR structure emerges as a broad (short lifetime) resonance centered at around 0.19 THz. This should give rise to the appearance of increased spectral amplitude in the corresponding frequency region and therefore, relatively large precession amplitude survives in the off-resonant temperatures below 84 K.

4.4.2 Creation of Macroscopic Magnetization by SRPT

Now, we proceed to the main experiment: control of the phase transition path of heating-induced SRPT with THz-induced spin precession. This time, as explained above, we turned on the optical heating pulse in addition to the THz pump.

The waveform (black trace) indicated in the top graph in Fig. 4.12 shows the transient Faraday rotation measured without applying the heating pulse, which is the same waveform at 84 K as in Fig. 4.11. The oscillation of M_z due to spin precession caused by magnetic nearfield of SRR can be clearly seen. In contrast, when we applied the heating pulse, the waveform behaved quite differently. The middle 2D image of Fig. 4.12 shows the dynamics of the out-of-plane magnetization M_z measured at 84 K for various values of the incident timing *dt* of the heating pulse with respect to the THz pulse. After the application of heating pulse, the Faraday rotation signal drastically increased and reached the final state with a large finite value that lasted for longer than several nanoseconds (on this color scale, the spin precession before application of heating pulse cannot be resolved because its amplitude is too small). Surprisingly, Faraday amplitude measured at the final state was nearly two orders of magnitude greater than the original spin precession amplitude. As we discuss in the next subsection, this value was over 60% of the value of Faraday rotation where total magnetization was aligned by external magnetic field along $\pm c$ direction.

Also, it can be seen clearly that the phase of created magnetization M_z in the final state oscillates against *dt*. Plot of its final values as a function of *dt* (graph on the right) clearly shows that the frequency and phase of this oscillation are similar to the original spin precession waveform. Hereafter we call this waveform "*dt*-waveform". To see this in more detail, typical transient Faraday dynamics measured for two fixed values of *dt* (= 57 and 65 ps) are shown in Fig. 4.13. The rise time after the heating pulse is shown to be around 20 to 30 ps, which agrees with the previously reported rate of energy transfer from the phonon to the 4*f*-electron system (i.e., thermalization rate of the spin system) [2] and indicates that the drastic increase of Faraday

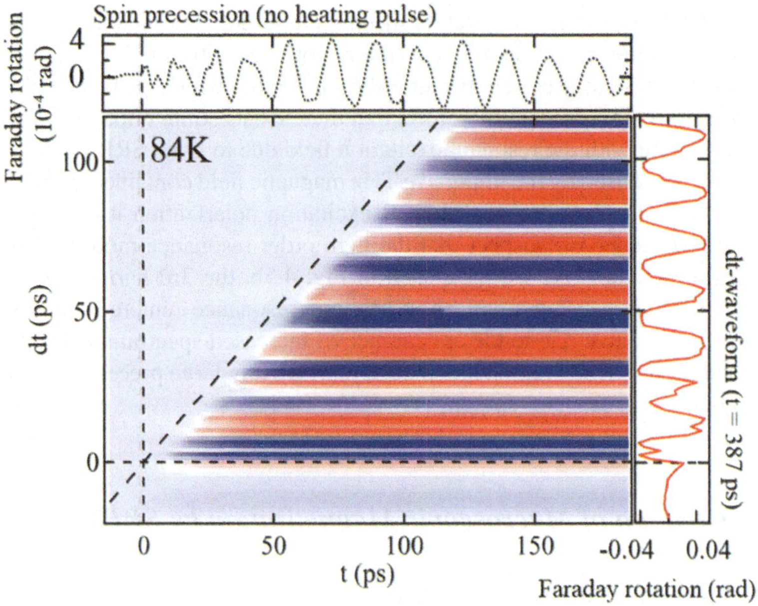

Fig. 4.12 Transient Faraday rotation waveform measured for various arrival timing delay (dt) between THz and heating pulse (middle 2D image), Faraday rotation waveform measured without heating pulse (graph on the top) and the final value of waveform as a function of *dt* (graph on the right), measured with the probe time fixed at a sufficiently later time $t = 387$ ps

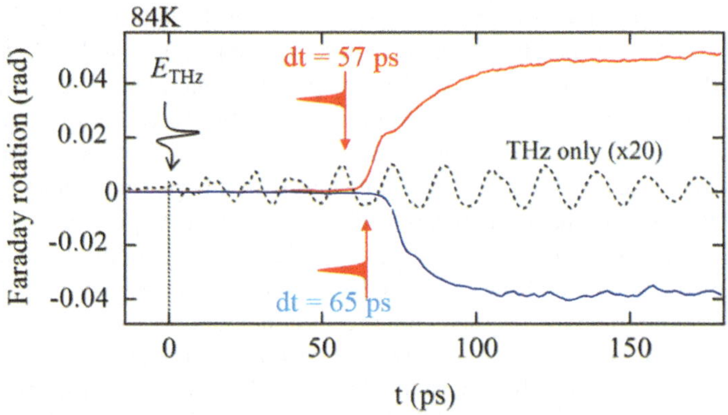

Fig. 4.13 Transient Faraday rotation signal measured without the heating pulse (black) and heating pulse incident at 57 ps (red)/65 ps (blue) after the THz pulse

rotation is caused by the rotation of magnetization towards out-of-plane direction due to the heating-induced SRPT. It is seen that when the heating pulse was applied at the moment the spin was slightly tilted towards + *c*-axis ($dt = 57$ ps), the created magnetization M_z was also directed towards the + *c* direction. Conversely, when the timing of heating pulse *dt* was changed to 65 ps when the spin was tilted to the –*c*-direction ($dt = 65$ ps), then the direction of created magnetization was reversed to the –*c* direction. This shows clearly that the direction of created magnetization is determined by the direction of initial tilting of spin system at the moment of irradiation by heating pulse. If we assume that the sample temperature reaches T_L immediately after the optical excitation, this result agree with the model we proposed previously in the Sect. 4.2 (Fig. 4.2) that the path of SRPT is determined by the instantaneous phase of the coherent spin precession at the moment the sample temperature has reached the SRPT temperature T_L.

Furthermore, in order to prove that the observed creation of macroscopic magnetization is actually caused by the magnetic nearfield of SRR, we measured the response inside/outside of the SRR structure. As shown previously in Fig. 4.7, the direction of the out-of-plane magnetic nearfield H_z reverses its sign at the opposite sides of the stripe, reflecting the current flowing in it. The resulted *dt*-waveforms are shown in Fig. 4.14. As expected, the direction of the created magnetization was

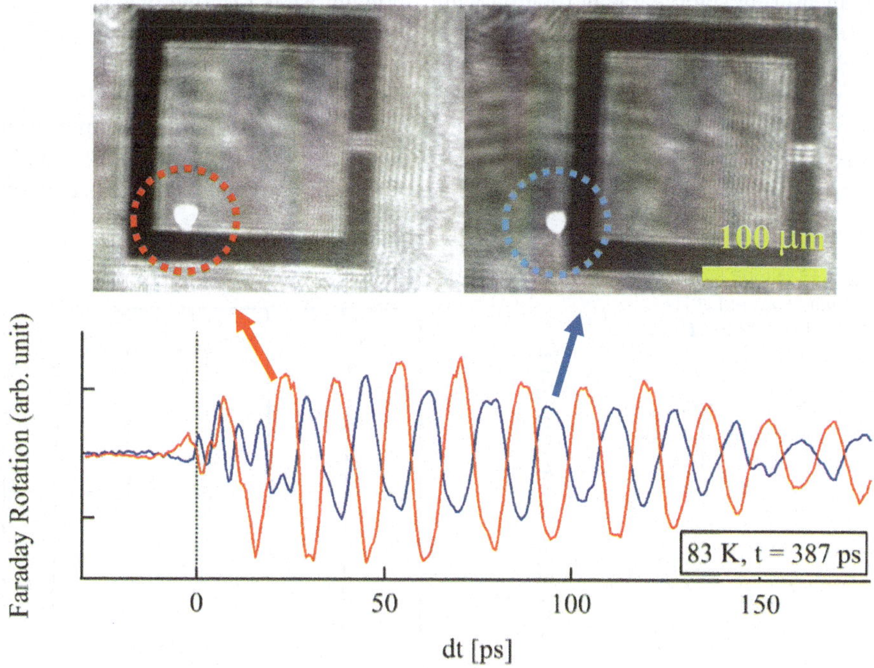

Fig. 4.14 The *dt*-waveforms measured inside (red) and outside (blue) of SRR structure. Insets are the microscope images of the same SRR showing corresponding probe spot positions (white spot near the left bottom). The contrast of the image are enhanced for clarity

completely reversed inside and outside of the SRR. This shows clearly that the observed phenomenon is induced by the *magnetic* nearfield of SRR structure. Also, this suggests that the creation of macroscopic magnetization was realized on a spatial resolution of approximately 30 μm, or 1/16 of the incident THz spot size (~500 μm). The spatial resolution of this magnetization formation is determined by the size of the metallic structure and also by the diameter of the optical pump spot (~100 μm in this case), which means that it may be scaled down to even smaller dimensions by proper designing of the optical focus systems and metallic structures.

4.4.3 Estimation of Tilt Angle of Spins

In this subsection, we give the quantitative evaluation of the macroscopic magnetization created in the THz-induced SRPT alignment process. For this, we have to know the value of Faraday rotation in the case where total magnetization in the sample was uniformly aligned along the out-of-plane direction. In order to do this, we measure the Faraday rotation under the HT phase.

The macroscopic magnetization of $ErFeO_3$ is directed towards $\pm c$ in the HT phase, which is along the out-of-plane direction for our *c*-plane cut sample. Therefore, by measuring static Faraday rotation while applying weak external magnetic fields along $\pm c$ direction in the HT phase, it is possible to evaluate the Faraday rotation angle for saturated magnetization domain along out-of-plane direction. Figure 4.15 shows the static Faraday rotation measured at $T = 100$ K, which is slightly above T_H. Here, the static magnetic field was applied by placing the permanent magnet from outside of the cryostat, and its direction was inverted every 10 s. The magnetic field strength reached approximately ± 500 Gauss inside the cryostat, which was enough

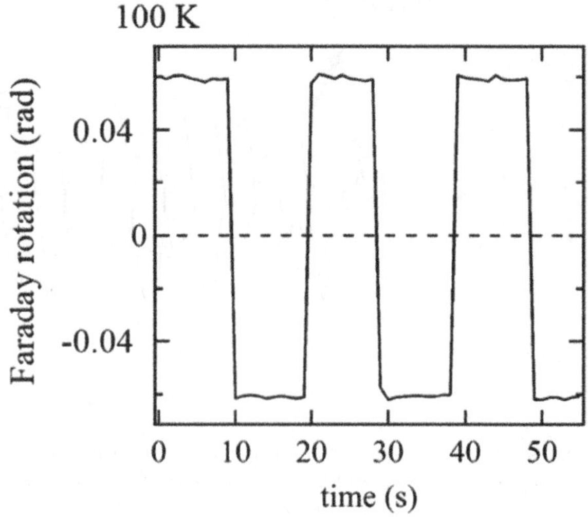

Fig. 4.15 Static Faraday rotation measured at HT phase (T = 100 K). The direction of external magnetic field (approximately 500 Gauss) was applied along ± *c*-axis and inverted every 10 s

to saturate magnetic domains along $\pm c$. It can be seen that the magnitude of Faraday rotation for saturated magnetization is $\pm$ 0.061 (rad).

By comparing this value with the measured transient Faraday rotation amplitudes, it is possible to evaluate the amount of out-of-plane magnetization component or the tilting angle.

For example, the maximum value of the out-of-plane magnetization created after THz / heating pulse excitation (see, e.g., the value at t = 170 ps of the blue curve in Fig. 4.13) was estimated to be 0.039 / 0.061 = 63% of the total magnetization. In other words, 82% of the magnetization in the sample is pointing -*c* and other 18% is directed towards + *c*.

Tilt angle of magnetization during spin precession can be also evaluated in a similar manner. The maximum amplitude of Faraday rotation caused by spin precession, such as shown in Fig. 4.11a, was approximately 4×10^{-4} (rad). Therefore, the average tilting angle during precession can be estimated to be $\Delta\theta = \sin^{-1}(4 \times 10^{-4}/0.06) \sim 0.4$ (deg).

4.4.4 Incident THz Amplitude-Dependence of the Created Magnetization

Next, in order to evaluate the amount of the created magnetization and to get insight into its formation mechanism, we measured its dependence on the incident THz amplitude. Here, we focus on the *dt*-waveform mentioned earlier. The incident THz amplitude was varied by rotating one of the wiregrid polarizer pair.

The *dt*-waveforms measured at 84 K for the incident THz amplitude ranging from E_{THz} = 300 kV/cm (100%) to 30 kV/cm (10%) are shown in Fig. 4.16a. While their frequency and phase are almost similar, the amplitude of the waveforms are clearly saturated under higher THz fields. For clarity, the normalized waveforms for 30 and 300 kV/cm are plotted in Fig. 4.16b. It can be seen that the waveform for 300 kV/cm exhibits clamped and rectangular-like shape. Typical dependence of the Faraday amplitudes as a function of incident THz field [Fig. 4.16c] shows this saturation behavior more clearly. The amplitude of the original spin precession without heating pump was almost linear under the incident THz amplitude, indicating that the saturation was not caused by the nonlinearity of the spin precession dynamics. The maximum value of Faraday signal at saturation [0.04 (rad)] indicates that approximately 63% of the total magnetization in the sample was aligned along $\pm$ *c* direction, as mentioned in the previous section. This estimation gives the lower limit, because the amount of magnetization should be larger in the region close to the front surface where there was sufficient heating.

In the weak THz amplitudes regions, in contrast, the dependence shows a linear response. Such crossover behavior from the linear to saturation behavior can be understood by taking into account a finite distribution of magnetization around its equilibrium position (Fig. 4.17), possibly originating from thermal fluctuation or

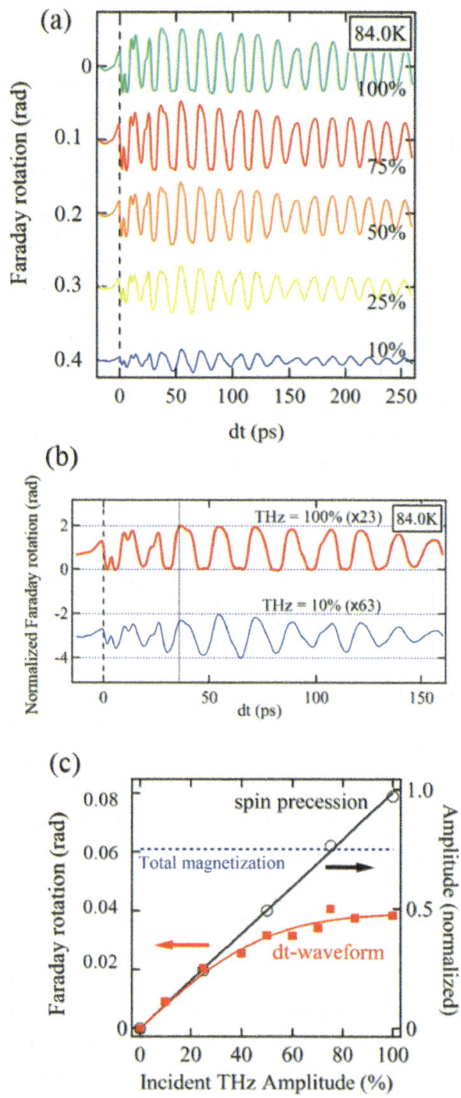

Fig. 4.16 **a** *dt*-waveforms measured inside the SRR for different strengths of incident THz amplitude. Here, 100% corresponds to 300 kV/cm in peak electric field. **b** The normalized *dt*-waveforms for the incident peak THz amplitudes of 300 kV/cm and 30 kV/cm. The waveform for the 300 kV/cm is clearly saturated. **c** THz amplitude-dependence of the traces in (**a**) evaluated as the average within 4 ps around $dt = 38$ ps (red filled square) and that of spin precession amplitude measured without heating pulse (black open circle), evaluated as the root-mean-square amplitude of the waveform in the time range $t = [-13 \text{ ps}, 253 \text{ ps}]$. Solid curve is the fit by an error function. Blue line indicates the amplitude of static Faraday rotation measured in the HT phase, with M_z completely aligned towards $+c$ or $-c$ directions by the application of external static magnetic fields

domain distribution in the sample, as discussed later. If we assume the existence of such finite distribution of spins, then the measured Faraday rotation is given by an ensemble average of such magnetic moments within the probed region. In the case of thermal fluctuation, for example, the shape of this distribution may be approximated by a Gaussian function. We can also assume from the results obtained in the former section that the amount of magnetization in the final state is given by the difference of the number of spins in the $+c$- and $-c$- sides at the moment of irradiation by a heating pulse.

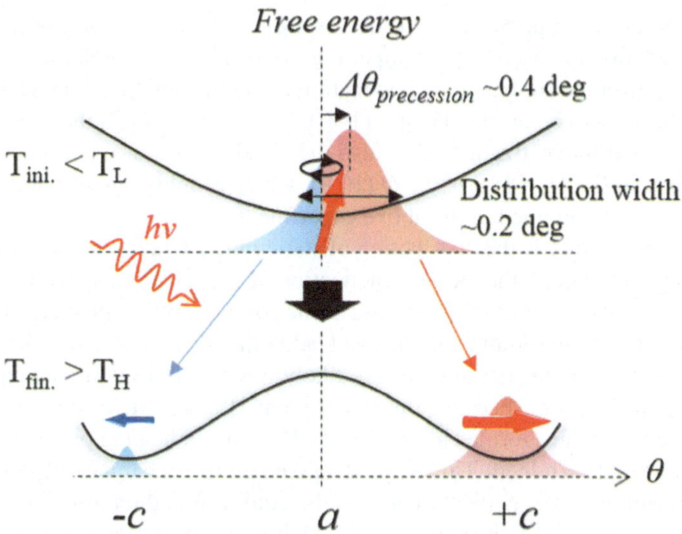

Fig. 4.17 Schematic model of the saturation behavior assuming a finite distribution of magnetic moments within a magnetic free energy potential. When the spin precession amplitude is small, the amount of the final-state magnetization is given by the small shift of population of the spin distribution and thus, THz-amplitude dependence is expected to exhibit a linear function. On the other hand, when the spin precession amplitude exceeds the width of this distribution, the resulted magnetization is completely aligned along $+c$ or $-c$ directions, thereby saturation occurs

When such ensemble of spins are excited by a weak THz magnetic field, the spin system is only weakly fluctuated and therefore, the number of spins at each sides change only linearly with respect to the amplitude of spin precession. On the other hand, under the high incident magnetic field strength the tilting angle of coherent spin precession induced by the THz exceeds this width of distribution. This leads to the complete weight shift towards $\pm c$ during the precession motion, resulting in the saturated magnetization after heating.

In the above model, the THz amplitude dependence of the created magnetization is expected to be expressed by the integral of Gaussian function, i.e., an error function. In Fig. 4.16c we plotted the fit of experimental data by this function, which nicely reproduces the observed dependence and thus validates the abovementioned mechanism that the direction and amount of the created magnetization is actually determined by a small shift of spin population at the initial moment of optical heating. As will be mentioned in the following sections, such picture is also validated from the measurement of the temperature- dependence of the *dt*-waveforms. The width of Gaussian can be estimated from the fitting curve of Fig. 4.16c: It can be seen that the saturation becomes apparent over the THz amplitude of ~ 40%, which corresponds to the tilt angle of original spin precession of ~0.2°. Therefore, the distribution width of the abovementioned ensemble of spins is estimated to be of the same order.

As for the origin of such distribution, two possibilities are considered. First possibility is the thermal fluctuation of the spin system caused by finite temperature of the sample. The Néel temperature of $ErFeO_3$ is T_N ~ 600 K. Considering this and the Brillouin function to describe the temperature-dependence of sublattice magnetization $\boldsymbol{S}_{1,2}$, the average sublattice magnetization along the antiferromagnetic vector at 84 K is estimated to be around $|\boldsymbol{S}_{1,2}|$ ~ (1–10^{-5}). Because of this thermal excitation of $\boldsymbol{S}_{1,2}$, the total magnetization $\boldsymbol{M}$ is also expected to exhibit a similar fluctuation of the order $\Delta|\boldsymbol{M}|$ ~ $(1 - |\boldsymbol{S}_{1,2}|)^{1/2}$ ~ O (10^{-3}), which roughly agrees the distribution width (~O (0.1°)) obtained in the above experiment.

Another possibility is the distribution of static magnetization domains within the probed region. Although the total magnetization $\boldsymbol{M}$ is basically aligned towards $\boldsymbol{M}$ // $+a$ or $\boldsymbol{M}$ // $-a$ at the initial state, the coexistence of these two oppositely magnetized states within the finite sample volume can lead to the formation of magnetic domain walls. Magnetization in such regions are likely tilted from the original equilibrium directions. Furthermore, because our experiment is a repetitive measurement, the shapes of such domains formed after every 1 kHz shots of heating pulse are expected to be randomly determined. This may also give rise to appearance of the observed finite distribution of magnetization around the equilibrium direction when averaged by the time constant of lock-in detection, which is around 1 s/points in this case.

To conclude, our results clearly showed that due to the excitation by THz pulse, magnetization domain that is saturated and almost totally aligned to either direction within the heated region could be created in the SRPT process. To the best of our knowledge, this was the first demonstration that the THz magnetic field was successfully used to control the magnetization of condensed-matter system in a macroscopic level.

4.4.5 Temperature-Dependence of the Created Magnetization

Lastly, in order to further examine the magnetization formation mechanism, we measured the *dt*-waveform at various temperatures (Fig. 4.18). As can be seen, the waveform changes strongly with temperature. The most significant change can be seen in their amplitudes. Plotted in Fig. 4.19 (a) is the amplitude of each waveform as a function of temperature, evaluated as its root mean square in the range t = [0, 260] ps. It can be divided into following three characteristic temperature regions: (i) lower temperatures from T = around 74 K to 83 K (yellow-colored region), where the amplitude is small and slowly increases as temperature rises, (ii) T = around 83–84 K, where the signal significantly increases (red-colored), and (iii) T $\gtrsim$ 84.5 K, where it sharply drops to zero (blue-colored). In the following we will discuss the origin of each features in more detail.

(i) The relatively small amplitude in the low initial-temperature regions can be assigned to the incomplete rotation of magnetization after heating. In these temperatures, the temperature rise caused by the excitation with heating pulse is insufficient to make the spin system reach high temperature phase (HT) in the final state, leaving the magnetization in the intermediate angles between θ

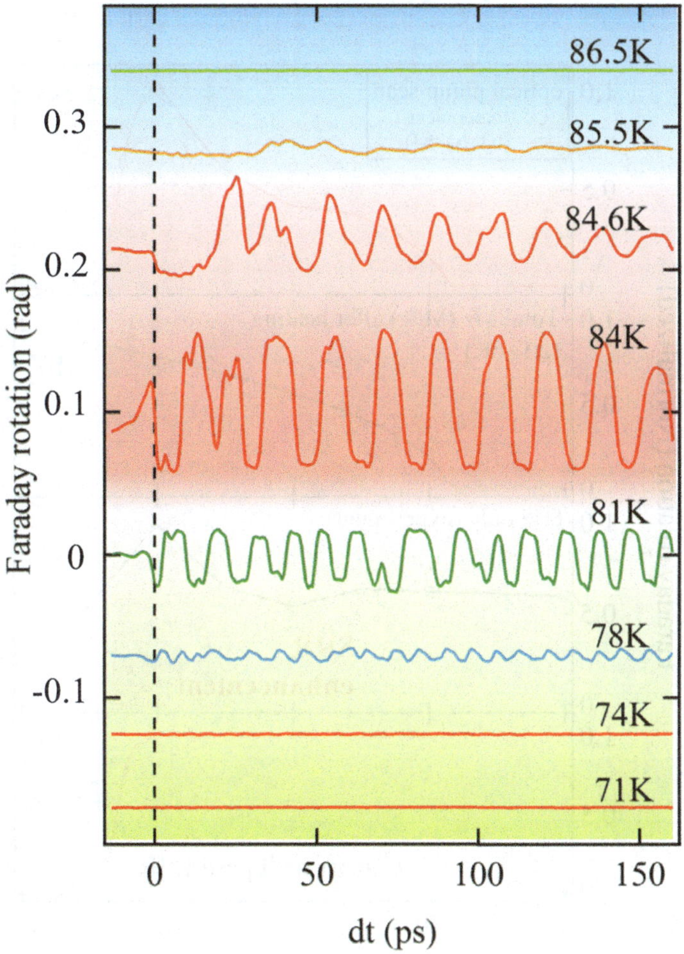

Fig. 4.18 *dt*-waveforms measured with maximum incident THz amplitude for various temperatures. The yellow-, red- and blue-colored regions correspond to the temperature regions which are dominated by different origins described in (i), (ii) and (iii) in the main text, respectively

= 0 ~ 90 deg. Figure 4.19b shows the final value of the out-of-plane magnetization component after heating as a function of initial temperature. This trace was calculated by shifting the trace of temperature-dependent rotation curve of magnetization near the SRPT region [Fig. 2.3c] by 6 K, i.e., the average temperature rise caused by heating was assumed to be $\Delta T = 6$ K. The trace qualitatively reproduces the tailed feature at low temperature region in Fig. 4.19 (a). Slight deviation may be due to the inhomogeneity of heating in the sample along the propagation direction of heating pulse.

(ii) The increase of signal amplitude at $T = 84$ K is assigned to the resonant enhancement of spin precession by SRR resonance. As mentioned previously,

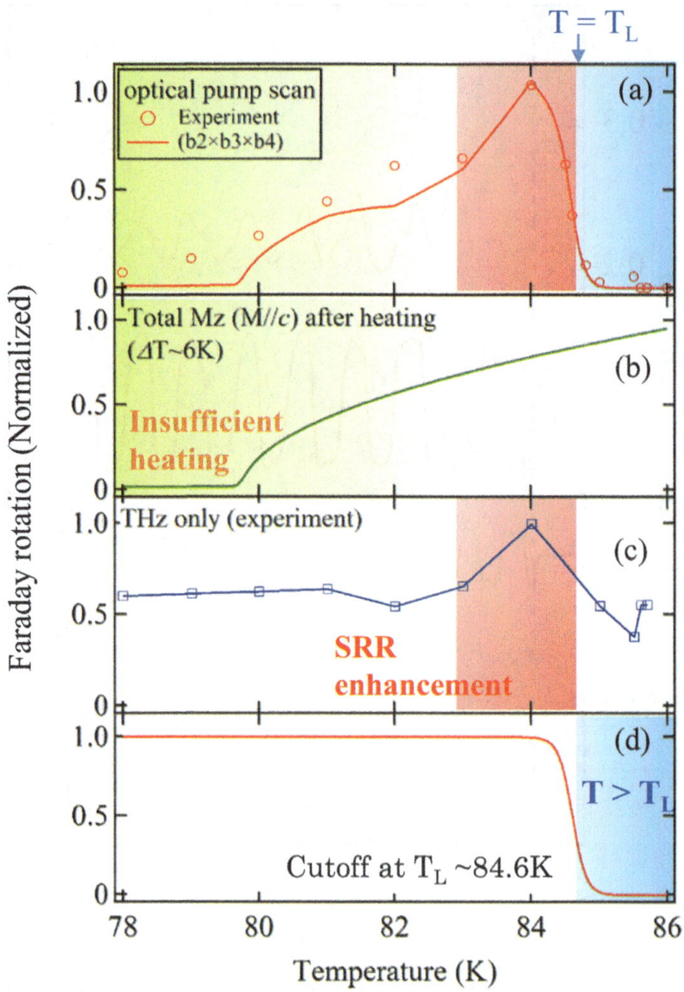

Fig. 4.19 Temperature dependence of the amplitude of *dt*-waveform (**a**) and the THz-only waveform (**c**). **b** The final value of the out-of-plane magnetization M_z assuming the temperature rise ΔT = 6 K, calculated by shifting the temperature-dependence curve of magnetization obtained from a procedure described in 2.1.3. **d** Cutoff behavior at T > T_L ~ 84.6 K approximated by an error function. Red solid curve in (**a**) was calculated by multiplying the three curves in (**b**)–(**d**)

the SRR structures were designed such that its resonance frequency (~20 GHz) matches the FM at 84 K. Figure 4.19c plots the amplitude of FM precession measured without heating pulse [original waveforms are shown in Fig. 4.19a] and therefore, the signal enhancement caused by resonant enhancement at this temperature is observed as mentioned previously.

(iii) The sharp decrease of signal occurs over a threshold temperature T_{th} ~84.5 K. This temperature coincides well with the lower temperature boundary of

SRPT T_L ~85 ± 1 K, estimated from the temperature dependence of the FM frequency in the previous section [Fig. 4.11c]. From this, we can presuppose that $T_{th} = T_L = 84.5$ K. Interestingly, the amplitude of original spin precession remains almost unchanged around this temperature [Fig. 4.19c], while only that of dt-waveform changes drastically. This indicates that the observed disappearance of the signal reflects some intrinsic nature of the magnetization formation process caused by SRPT.

In order to explain this behavior, we have to take into account the static tilting of the equilibrium angle of magnetization within the SRPT temperature region. As mentioned earlier, the SRPT of $ErFeO_3$ is a second-order phase transition and thus, the easy-axis of magnetization continuously rotates within the *a*–*c* plane. Therefore, over T_L, the magnetization points slightly away from the *a*-axis.

When the sample temperature is below T_L (LT phase), as is the case for yellow- and red- colored regions in Figs. 4.18 and 4.19, the equilibrium direction of magnetization is pointed towards *a*-axis and the magnetization precesses around this direction with an amplitude of ~0.4 degree, as shown in Fig. 4.20a. As long as the THz amplitude is large enough, the created magnetization after heating points towards either the + *c* or –*c* directions almost completely in the heated region, as mentioned in the former section.

On the other hand, when the initial temperature was set above T_L, the equilibrium direction of magnetization is slightly deviated from the *a*-axis [Fig. 4.20b]. Quantitatively, it can be seen from the temperature-dependence curve of magnetization shown in Fig. 2.3c that an increase of the temperature by 0.1 K immediately above T_L results in the rotation of the easy-axis direction of approximately 1 degree, which is of the same order with the spin precession amplitude (~0.4°, see Fig. 4.17).

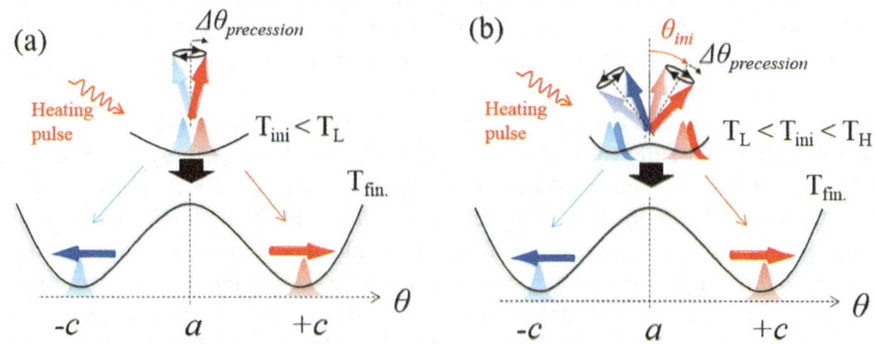

Fig. 4.20 Schematic explanation of the disappearance of signals in $T > T_L$ region due to static deviation of magnetization direction from *a*-axis. **a** When the initial temperature $T_{ini} < T_L$, as long as the amplitude of coherent spin precession induced by THz magnetic nearfield is large enough, the magnetization in the final state can be selectively formed. However, when $T_{ini} > T_L$, **b** the static tilting of the magnetization away from the original *a*-axis will overcome this precession amplitude and therefore, the population ratio towards ± *c* –axis cannot be modulated by the precession any more

Therefore, over this temperature the angle of static deviation of the spin system exceed that of the THz-induced precession. This means that the direction of final-state magnetization is determined only by the initial tilting direction of magnetization and cannot be influenced by the THz-induced precession motion. As mentioned in the first section, during the SRPT the rotation of the spin system towards $+ c$ and $-c$ directions occurs with an equal probability. Therefore, under such conditions, the amount of the created net magnetization is zero. Thus, the observed disappearance of signal above T_L was explained qualitatively. If we assume again that the distribution of magnetic moments has a finite width within potential minima (e.g., Gaussian function) and that its average position shifts linearly towards $\pm c$ direction with an increase of temperature, then the observed disappearance feature is expected to be roughly approximated by an error function, as shown in Fig. 4.19d.

By taking into account all the above mentioned effects (i)–(iii), we multiplied all the traces in (b), (c) and (d) to get the solid traces in Fig. 4.19a, which nicely reproduced the observed data throughout all temperature regions including above T_L. It is worth noting that the observed disappearance above T_L clearly verifies our assumption in the first place that symmetry breaking is caused by the small tilting of magnetization about equilibrium position in the initial process of SRPT.

4.4.6 Contribution of SRR Magnetic Fields on the Process of Macroscopic Magnetization Formation

So far, we could successfully understand the overall behavior of the macroscopic magnetization creation process within a simple assumption that the direction of created magnetization is the same as that of the initial tilting at the instance of phase transition. In this subsection, we further examine this hypothesis by comparing the original spin precession dynamics measured without heating pulse (t-waveform) with the temporal profile of resulted magnetization (dt-waveform).

Figure 4.21a shows the waveforms measured at 80, 81, 82 and 84 K. The black and red traces are the t-and dt-waveforms, respectively. In every temperatures except for 84 K, the original spin precession (t-waveform) decays within ~200 ps after excitation by THz pulse. If only the tilt of magnetization causes the formation of macroscopic magnetization after optical heating, no signal should appear in the dt-waveform after 200 ps. Interestingly, however, strong signal can be seen in the dt-waveform in the temporal region even after 200 ps where the spin precession has decayed. More importantly, the period of oscillation of this signal is approximately 16 ps (0.06 THz), which can be assigned to the SRR-resonance frequency.

To see this more clearly, the waveforms measured at 81 K is expanded in the temporal region from 110 to 280 ps in Fig. 4.21b. Clearly it can be seen that around the time where there is sufficient amplitude in the spin precession (see, for example, t = 110–170 ps in the t-waveform), the resulted magnetization (dt-waveform) follows the frequency of original spin precession motion. The phase delay between t- and

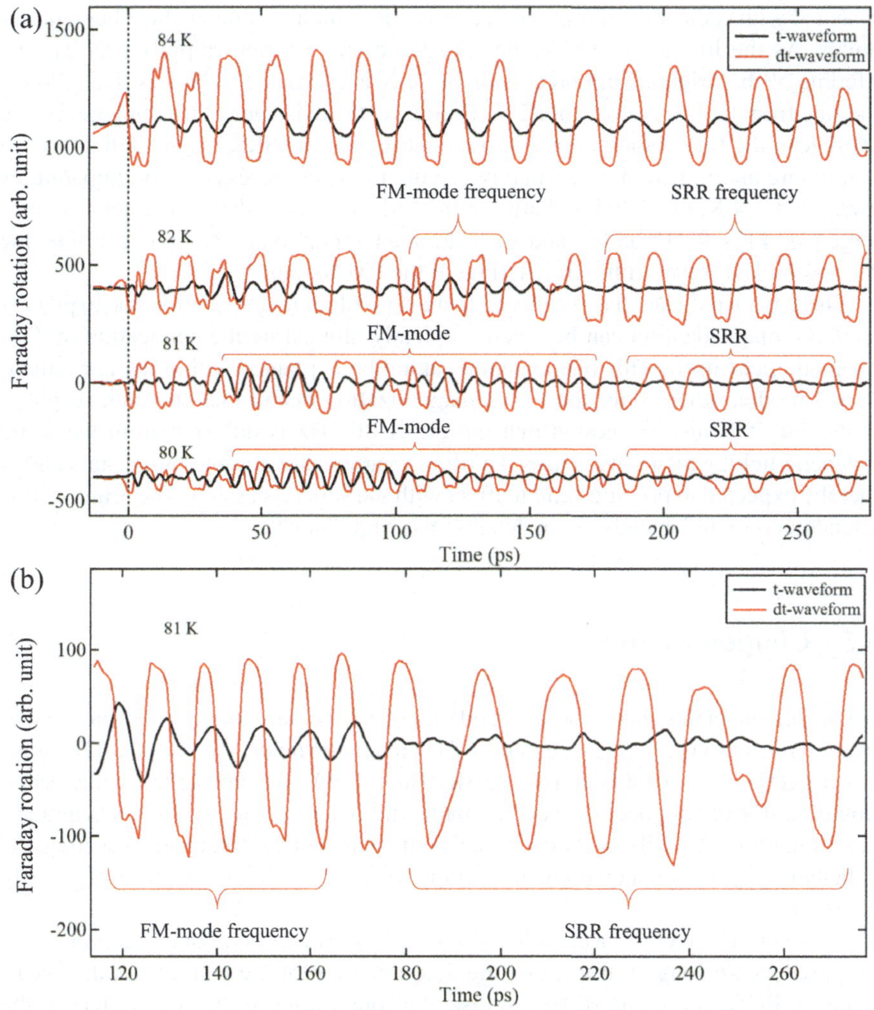

Fig. 4.21 **a** The t- and dt-waveforms measured at off-resonant temperatures (T = 80, 81 and 82 K). The dt-waveform follows the motion of t-waveform (original spin precession) in the regions where there are sufficient amplitudes of spin precession, however, in the small-amplitude regions it exhibits an oscillation with the period of SRR-resonance. **b** The waveforms at 81 K expanded within 110–280 ps

dt-waveforms during this time is ascribed to the time it is need for the spin system to reach T_L after optical excitation. As time progressed and the precession has decayed after t = 200 ps, the dt-waveform keeps a similar amplitude with the former region but its frequency is changed to that of the SRR resonance.

It should be noted that the SRR magnetic field itself is still present in this time region. In the t-waveform of 84 K where the FM-mode and SRR frequencies are matched with each other, the spin precession still remains unlike in the other temperatures. As the lifetime of FM-mode spin precession shortens rapidly near T_L, the lifetime of the original spin precession at 84 K is estimated to be less than 200 ps, judging from the t-waveform of 82 K. Therefore, the existence of the spin precession at 84 K in the time region after 200 ps suggests that the SRR magnetic field is still continuing and this field is resonantly driving the spin precession. At temperatures away from 84 K, the detuning between the FM-mode and SRR frequency becomes large [see Figs. 4.11b and c] and such resonant excitation does not occur, thus the precession lasts only within the lifetime of the FM-mode.

These facts indicate that even when there is no tilt of magnetization due to precession, the magnetization can be macroscopically aligned by the application of THz magnetic *field* itself. This phenomenon cannot be explained within the abovementioned model, where only the tilt of magnetization determines the path of phase transition. It is also suggested that the effect of THz is different from the static magnetic field, which only causes the tilt of magnetization in the initial state and is thereby expected to provide similar effect with the spin precession. The origin of this phenomenon will be analyzed in detail in the next chapter.

4.5 Chapter Summary

In conclusion, in this chapter we realized the coherent control of macroscopic magnetization formation process using intense THz magnetic fields. We used the THz fields as a trigger of the optical heat-assisted SRPT in $ErFeO_3$, which resulted in the formation of uniformly aligned magnetic domain with macroscopically broken symmetry in the final state. By utilizing two separate pump pulses of THz and optical heating, the coherent dynamics of such domain creation process was clearly and unambiguously revealed.

The overall behavior can be well understood by the model similar to that previously proposed by de Jong et al. [1] that the determination of the path of ultrafast heat-induced SRPT is dominated by the instantaneous tilting of the spin system at the moment the sample temperature has reached SRPT region T_L. However, by closely comparing the original spin precession (t-waveform) dynamics with the temporal profile of the resulted magnetization (*dt*-waveform), it was indicated that the transient THz magnetic *field* itself produced by the SRR also contributes strongly to the domain formation even when the actual tilting of magnetization at the initial process is small enough, which cannot be explained within the framework of previous model. In the following Chap. 5 we examine this phenomenon in more detail, aiming at obtaining the physical picture of its underlying mechanism.

References

1. de Jong JA, Razdolski I, Kalashnikova AM, Pisarev RV, Balbashov AM, Kirilyuk A, Rasing Th, Kimel AV (2012) Coherent control of the route of an ultrafast magnetic phase transition via low-amplitude spin precession. Phys Rev Lett 108:157601
2. Yamaguchi K, Kurihara T, Minami Y, Nakajima M, Suemoto T (2013) Terahertz time-domain observation of spin reorientation in orthoferrite $ErFeO_3$ through magnetic free induction decay. Phys Rev Lett 110:137204

Chapter 5
Numerical Simulation of the Macroscopic Domain Formation

Abstract The aim of this chapter is to clarify the mechanism of macroscopic symmetry breaking process observed in Chap. 4. The important results to be explained are summarized: (1) Irradiation of optical heating pulse can induce the SRPT, causing the magnetization $\boldsymbol{M}$ to rotate from $\boldsymbol{M} \,//\, a$ to $\boldsymbol{M} \,//\, \pm c$, (2) By exciting the system with THz magnetic field in advance to the heating pulse, one of the two (+*c* or –*c*-aligned) macroscopically magnetized domain states can be selected in the final state, reflecting the transient phase of spin precession at the moment of photoirradiation, (3) THz amplitude-dependence measurement suggests that the direction of $\boldsymbol{M}$ has a finite distribution around equilibrium direction, (4) Even when the spin precession amplitude is small (or zero), macroscopically magnetized domain can be created, reflecting the amplitude and direction of the THz magnetic field applied by the SRR structure. The result (4) cannot be explained under the conventional model that the tilting direction of magnetization at the moment of heating determines the resultant magnetization state. The main aim of this chapter is to give a microscopic explanation to this phenomenon. We do this by simulating the spin dynamics using Landau-Lifshitz-Gilbert (LLG) equation, which describes the classical motion of the spin system. We first give an overview of the model and then fix the values of essential parameters quantitatively. Finally, using such model we aim to identify the physical mechanism responsible for the creation of macroscopic magnetization domains induced by THz fields.

Keywords Landau-Lifshitz-Gilbert equation · Free energy · Magnetic anisotropy

5.1 Simulation of Magnetization Dynamics Using LLG Equation and Free Energy Model

5.1.1 2-Spin Free-Energy Model

As mentioned in Chap. 2, the static and dynamical behavior of the $ErFeO_3$ magnetization are known to be well described by the LLG equation along with the 2-spin free energy model described by Eq. (2.1).

T. Kurihara, *Observation and Control of Magnetic Order Dynamics by Terahertz Magnetic Nearfield*, Springer Theses, https://doi.org/10.1007/978-981-16-8793-8_5

In this model, the behaviors of equilibrium position and the magnetic resonance modes are dominated by the second order anisotropy terms A_{xx} and A_{zz}. For $ErFeO_3$ it is known from the temperature dependence of the magnetic resonance frequencies that the parameter A_{zz} is the most predominantly temperature dependent term. In the following we fixed the values of other temperature-independent terms as $J = 20$, $\boldsymbol{D} = (0, -0.86, 0)$, $A_{xx} = -0.1$ and $A_4 = 0.0007$ (all units in cm^{-1}), based on the parameters derived in the previous researches [113, 114].

5.1.2 Equilibrium Spin States

The equilibrium states of the spin system can be calculated from eqs. (2.1) and (2.7)–(2.10) by assuming $\boldsymbol{H}_{THz} = 0$ and that the left hand sides of the eqs. 2.7 and 2.8 are zero. In this way, the shapes of the free energy curve and equilibrium states of the spins are calculated. Figure 5.1a shows the free energy curves at the temperatures of 74 K (LT), 88 K (during SRPT) and 106 K (HT). It can be seen that the free energy exhibits single minimum around $\boldsymbol{M} \mathbin{//} a$ in the LT, double minimum in the SRPT temperature region and minimum points around $\boldsymbol{M} \mathbin{//} \pm c$ in the HT phases. Correspondingly, it can be clearly seen that the spin system rotates continuously within the x–z plane with temperature as can be seen in Fig. 5.1b.

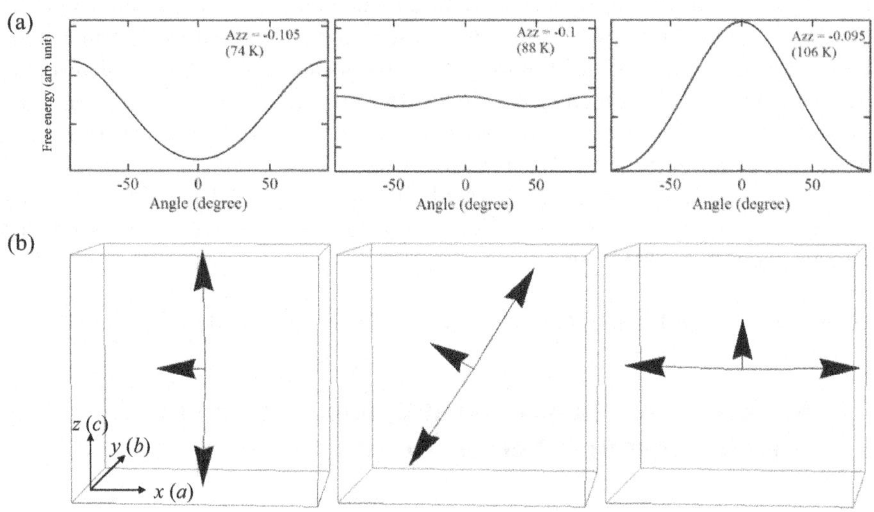

Fig. 5.1 **a** Shapes of free energy curve near the SRPT temperature region, calculated from Eqs. 5.1 and 5.2. **b** Equilibrium states of the two sublattice spins $\boldsymbol{S}_1$ and $\boldsymbol{S}_2$, and the total magnetic moment $\boldsymbol{M} = \boldsymbol{S}_1 + \boldsymbol{S}_2$. Here, $\boldsymbol{M}$ is exaggerated by 10 times for clarity. Here, the temperature was varied by changing the parameter A_{zz} using the relation shown in Fig. 5.2b

5.1.3 Temperature-Dependence of Spin Resonance Frequency

Next, we derive the relationship between A_{zz} used in the simulation and the absolute temperature T of the sample during experiment. As mentioned previously, the frequencies of the FM- and AFM modes are determined by the parameter A_{zz}. Using this, by comparing the frequencies of the calculated resonance modes at each values of A_{zz} with the experimentally obtained resonance frequencies at each temperature, it is possible to identify the relationship between these values. We simulated the spin dynamics under the condition that the spins $\boldsymbol{S}_1$ and $\boldsymbol{S}_2$ are excited by an impulsive magnetic field polarized along [111] direction and plotted the frequencies of subsequent oscillation corresponding to the precession motion of the total magnetization $\boldsymbol{M}$ (FM-mode) and the fluctuation of the length of $\boldsymbol{M}$ (AFM-mode).

Figure 5.2a shows the calculated temperature-dependence of the two resonance modes. Here, we approximated the temperature-dependence of $A_{zz}(T)$ by a following inverse-proportional function (Curie–Weiss law):

$$A_{zz}(T) = -a/(T + b) + C.$$

This form originates from the fact that the anisotropy field of Fe^{3+} spin systems is predominantly determined by the paramagnetic magnetization of 4f electrons in Er^{3+} ions [1]. We have chosen the parameters a, b and C so that the overall shapes of the temperature-dependence curves of FM- and AFM-mode frequencies best reproduced the experimentally obtained curves (Fig. 2.5 [2]). Figure 5.2b shows the reproduced curve of A_{zz} (T) near the SRPT temperature regions. It can be seen that around 80 to

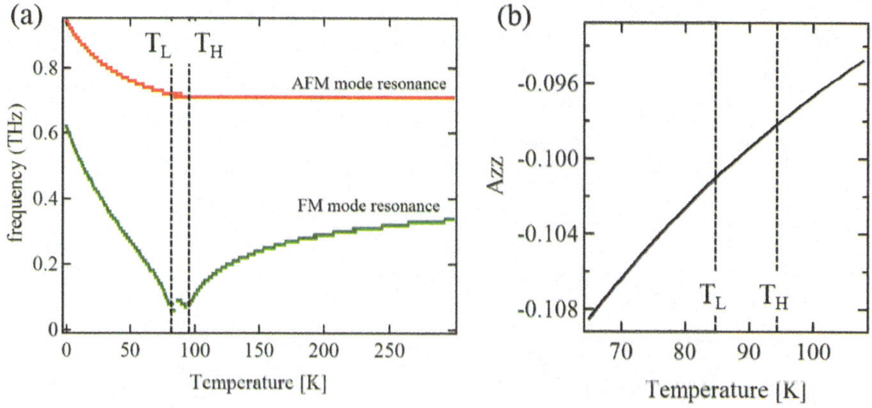

Fig. 5.2 **a** Temperature-dependence of the two magnetic resonance modes (FM- and AFM-modes) reproduced in the model calculation. **b** Relationship between the sample temperature and anisotropy parameter A_{zz} (T). The lower (T_L) and higher boundaries (T_H) of the SRPT temperature region is indicated by vertical dotted lines

100 K, A_{zz} (T) is essentially linear against temperature. Therefore, in the following we approximate A_{zz} (T) to be a linear function.

5.1.4 Spin Precession Excited by SRR Magnetic Fields

Using the parameters determined in the abovementioned procedures, next we calculated the spin precession dynamics after excitation by SRR magnetic fields. The THz waveform used in the calculation is shown as the black curve in Fig. 5.3a. Base on the result of FDTD simulation in the Sect. 4.3.1, it was designed to consist of two frequency components: the long-duration (>300 ps) oscillation with 0.06 THz frequency (main resonance mode of SRR) and the short-duration (< 20 ps) component centered at 0.19 THz (3rd harmonics mode, see Fig. 4.5b). The amplitude ratio of the two peaks was adjusted so that the ratio of SRR and FM-mode peaks in the simulated spectra of spin precession reproduced the observation [Fig. 4.11b]. The decay constant of spin precession α was also adjusted for each temperature to reproduce the experimental waveforms as shown in Fig. 5.3c. It is divergent near SRPT temperature T_L, which is the same tendency reported previously in the THz-TDS measurement [2].

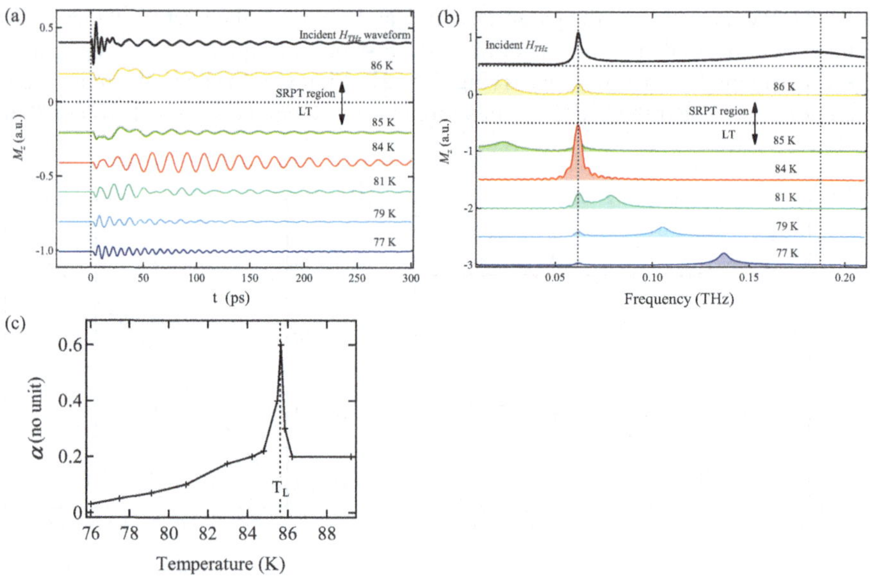

Fig. 5.3 **a** Time evolution of the z-component of the total magnetization $\boldsymbol{M}$ (t) calculated at each temperature. Black curve shows the incident THz magnetic fields. **b** Fourier spectra of (**a**). **c** Decay rate of spin precession α used in the calculation

The resulted waveforms shown in Fig. 5.3a well reproduced the main features observed in the experiment. It can be seen that the precession was resonantly enhanced at 84 K due to SRR. Below this temperature the precession exhibits a shortened lifetime and the beating structure caused by detuning between SRR and FM-mode frequency. At temperatures sufficiently far from this resonance temperature, the FM-mode still exhibits comparable amplitude in the early stage of precession (see, e.g., waveform at 77 K), because of the excitation by 3rd harmonics mode-component of incident THz field.

5.1.5 Simulation of Ultrafast Heating Process

As the next step we regard A_{zz} as a time-dependent parameter in order to express ultrafast heating of the spin system. The time evolution of A_{zz} was approximated by an exponential function, as shown in Fig. 5.4a. Here, as a typical case, the initial

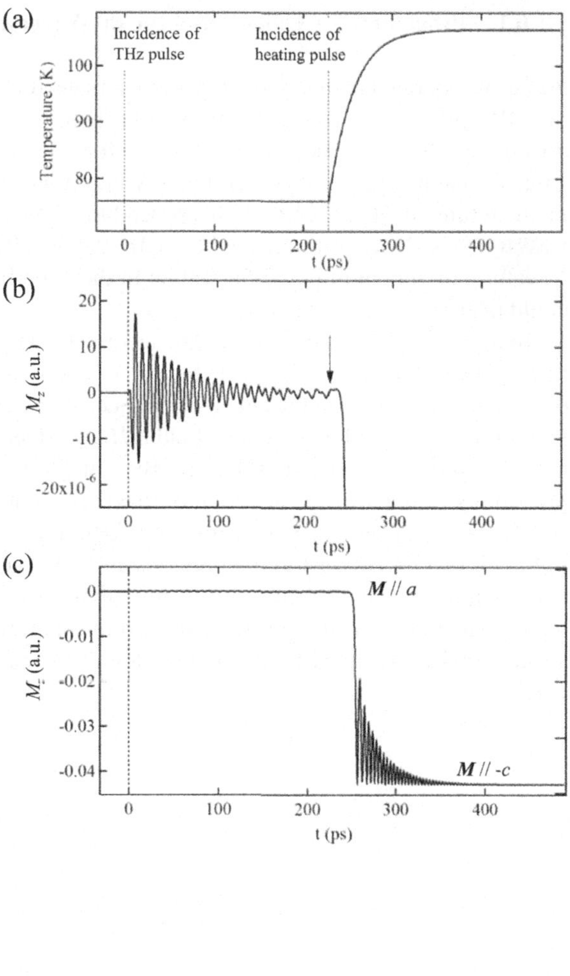

Fig. 5.4 **a** Typical time evolution of the anisotropy parameter A_{zz} (t) used in the simulation. Here, $T_{\text{ini}} =$ 76 K and the heating pulse is incident at $t = 228$ ps. **b** Calculated time evolution of the z-component of the total magnetization $\boldsymbol{M}$ (t) shown in **c** expanded around the a-axis. **c** Overall dynamics of $\boldsymbol{M}$ (t)

temperature is set at 76 K. The temperature rise induced by the irradiation of heating pulse is assumed to be 30 K, which is enough to cause complete rotation of the spin system from $\boldsymbol{M}$ // a to $\boldsymbol{M}$ // $\pm$ c after the SRPT if the initial temperature is in the range $T = 74$–85 K.

The time evolution of the z-component of total magnetic moment M_z is plotted in Fig. 5.4b and c. After the incidence of the THz magnetic field at $t = 0$ ps, the magnetization oscillates around the a-axis [see Fig. 5.4b], indicating precession motion of FM-mode. After the incidence of heating pulse at $t = 228$ ps, sample temperature gradually increases. As it reached $T_L = 86$ K at $t = 240$ ps, M_z rapidly drops within ~20 ps and settled into the final state with M_z // $-c$ (Fig. 5.4c). This behavior well reproduces the spin dynamics measured previously in Chap. 4 (see Fig. 4.12). Therefore, it was shown that the basic characteristic of the spin dynamics and the ultrafast heating-induced SRPT process can be successfully reproduced within this model.

5.1.6 *Simulation of dt-Waveform*

5.1.6.1 Procedure for Calculating the dt-Waveform

So far we obtained all the realistic parameters needed to describe the spin dynamics in SRPT process. In the next step we use them to calculate the dt-waveforms. To do this, we have to calculate the "t-waveform" similar to Fig. 5.4b and plot the value of magnetization M_z at sufficiently later time ($t = 360$ ps) for each values of incident time dt. However, with this procedure alone the resulted magnetization (dt-waveform) is always saturated either at M_z // $+$ c or M_z // $-c$ and exhibits a clipped waveform as shown in Fig. 5.5a, unlike in the experimental waveforms where they could take intermediate values.

To resolve this, we have to take into account the fact that the equilibrium direction of magnetization has a finite distribution around potential minima probably due to thermal fluctuation, as indicated in former Sect. 4.4.4. Here we mimic such fluctuation in terms of a tiny random magnetic field $\Delta\boldsymbol{H}_{\text{rand.}}$ added to the last term of Eq. 2.1, and take the averaged results of 100 times. By doing this, we can calculate the statistical probability of the *average* direction of magnetization rotation upon SRPT.

The dt-waveform calculated for the temperature of 84 K in this way is shown in Fig. 5.5(b). In contrast to the previous case, it is seen that the magnetization can take continuous values from $-c$ to $+c$ directions, in a similar manner with the experimental waveforms. Thus, the effect of the finite distribution of magnetic moment was successfully reproduced by phenomenologically including a randomly fluctuating magnetic field.

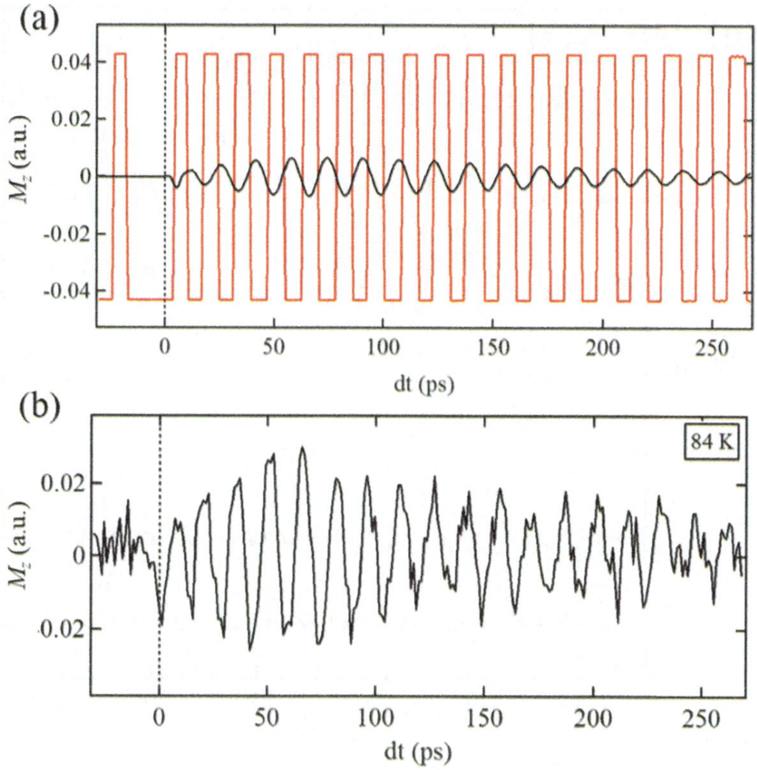

Fig. 5.5 The *dt*-waveform for 84 K calculated without (**a**) and with (**b**) the fluctuating magnetic field $\Delta \boldsymbol{H}_{\text{rand}}$. The black curve in (**a**) indicates the original spin precession without heating

5.1.6.2 THz Amplitude Dependence

Using the model, we checked if the essential behaviors of the *dt*-waveform observed in the experiments can be reproduced. For this, first we calculated the incident THz amplitude-dependence of the *dt*-waveform. Figure 5.6a shows the waveform at 84 K for various THz amplitudes. Clearly, the waveform saturates at higher incident THz fields and exhibits clamped shape. Typical amplitude-dependence at a fixed value of $dt = 51$ ps is plotted in (b), which nicely reproduces the tendency observed in Fig. 4.16a–c in the previous chapter. This indicates that the amount of created macroscopic magnetization is actually determined by the balance between the precession amplitude and finite distribution of magnetization, which is modeled by $\Delta \boldsymbol{H}_{\text{rand.}}$ in this calculation.

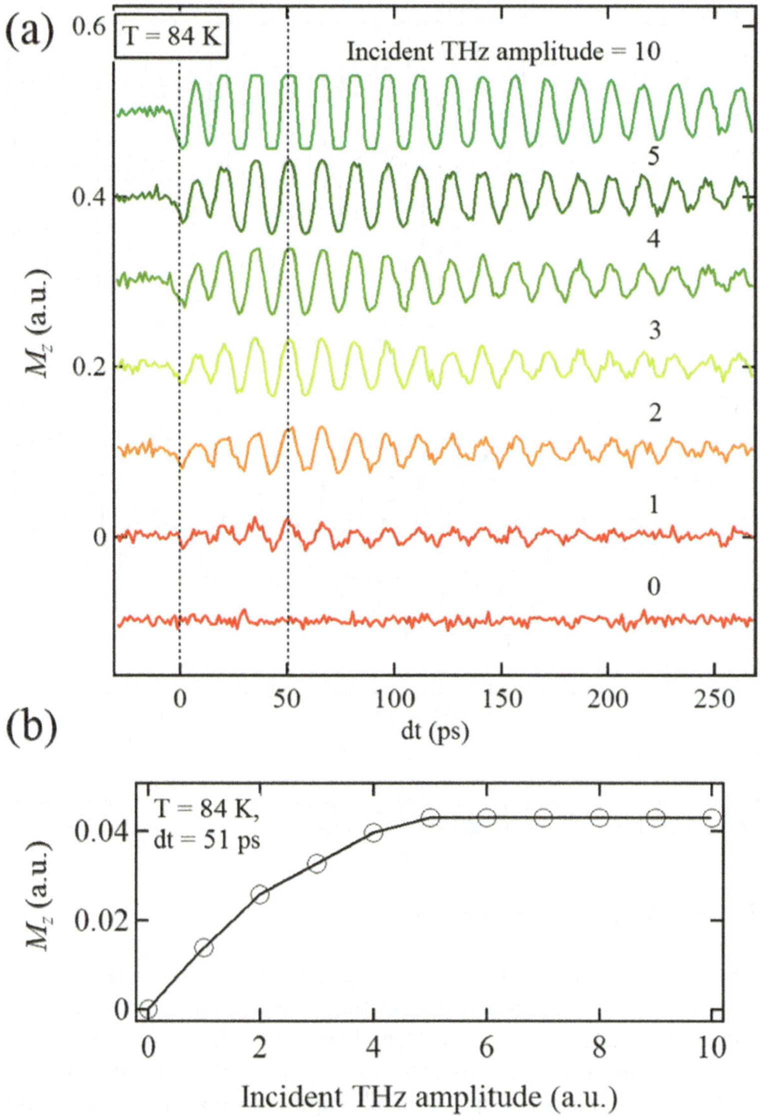

Fig. 5.6 **a** Calculated THz amplitude-dependence of the *dt*-waveform at 84 K. **b** Amplitude at *dt* = 51 ps plotted as a function of THz amplitude

5.1.6.3 Temperature Dependence

We also calculated the temperature-dependence of the *dt*-waveforms as shown in Fig. 5.7a. The corresponding amplitudes are plotted as a function of temperature in (b). They reproduced the main features observed in the experiment (Figs. 4.18 and

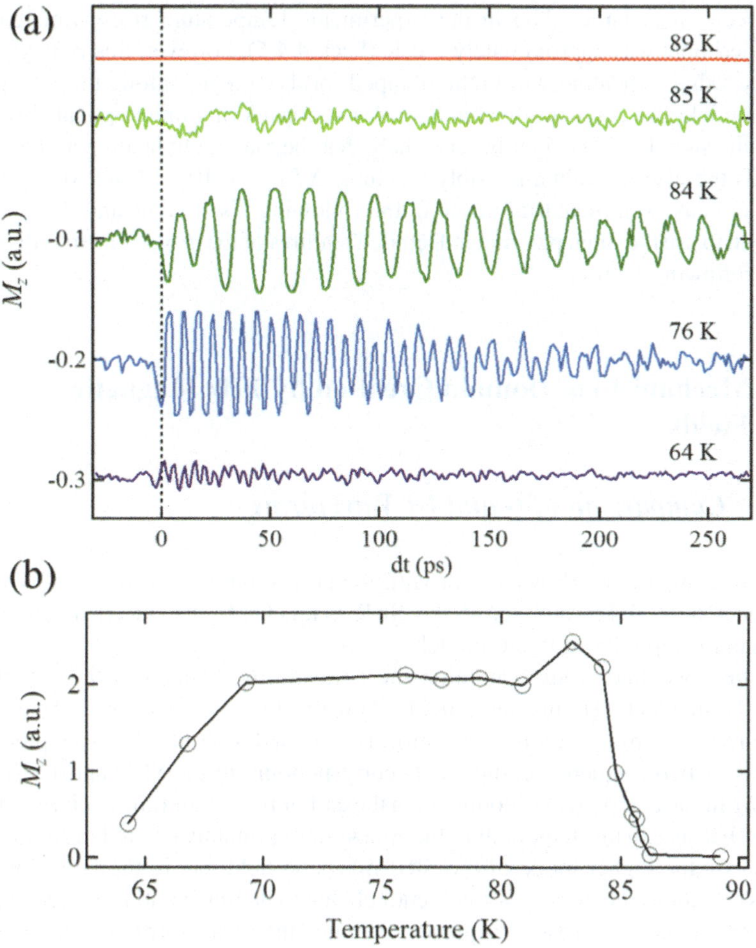

Fig. 5.7 **a** Calculated temperature-dependence of the *dt*-waveform. **b** Root mean-square amplitudes (within the time region $dt = 0$ to 268 ps) of **a** plotted as a function of temperature

4.19): (i) in the lower temperature regions (T < 70 K) the signal amplitude continuously decreases due to incomplete heating, (ii) around 84 K where the frequency of SRR and FM-mode precession were matched, it exhibits a peak due to the resonance enhancement of the spin precession amplitude and (iii) over T_L the signal rapidly decreases and completely disappears. Especially, the presence of (iii) in our calculation result clearly suggests that our aforementioned model is valid, that the signal disappearance in this temperature region is ascribed to the static rotation of the magnetization away from the original *a*-axis due to SRPT.

We can see in (b) a plateau structure in the temperature region of 70 K ~ 80 K, which was not so clear in the experimental results [Fig. 4.19a]. This is a consequence of the fact that the temperature rise which was set to 30 K in our calculation may have been much larger than in the experiment. Temperature rise estimated from the experiment was approximately ~ 6 K (Sect. 4.4.5), however, this only gives the average value of heating within the pumped (probed) region along the direction of sample thickness. As mentioned in Sect. 4.4.4, considering finite optical absorption inside the sample (OD ~ 1) it is very likely that the maximum heating at the surface exceeds the average value, possibly reaching $\Delta T_{max.} = 10 \sim 20$ K. At any rate, it does not cause significant influence in the following discussions and thus, we will continue to use the same parameter. Detailed features of the *dt*-waveform is discussed in the following section.

5.2 Mechanism of Domain Creation by THz Magnetic Fields

5.2.1 *Comparison of t- and Dt-Waveforms*

Here we compare the *dt*-waveform with the corresponding *t*-waveform as we did in former Sect. 4.4.5, and see if the SRR magnetic *field*-induced magnetization alignment is reproduced in our model.

At first, we start from examining the *t*-waveform obtained in the calculation. Here, the incident THz magnetic fields shown in Fig. 5.8a was used. As a typical case, in Fig. 5.8(b) we plot the t-waveform calculated for 76 K. The spin precession consists of two frequency components corresponding to the SRR and FM-modes. The FM-mode component is dominant in the earlier periods of time evolution ($0 < t < 200$ ps) because in this temperature the spin system is mainly excited in an impulsive manner by the higher mode of the SRR magnetic field, but in the later times ($t >$ 200 ps) its amplitude becomes small enough due to damping and is exceeded by the weaker oscillation with SRR frequency. We fitted the *t*-waveform with the following function:

$$M_z(t) = A_1 \sin(\omega_1 t - \phi_1) \exp\left(-\frac{t - t_0}{\alpha_1}\right) + A_2 \sin(\omega_2 t - \phi_2) \tag{5.1}$$

As shown by the black dotted curve of Fig. 5.8b, the waveform can be excellently well fitted by the above function. The envelopes for the ω_1- and ω_2-components ($A_1 \exp[-(t - t_0)/\alpha_1]$ and A_2) are also indicated in Fig. 5.8b as the red curve and the black line. It can be seen that the amplitude of ω_2-component (SRR frequency-component) is almost negligible compared to the ω_1-component (FM-mode).

The *dt*-waveform calculated for the same temperature is shown in Fig. 5.8c. In contrast to the previous case of *t*-waveform, the FM-mode frequency component is dominant in the first half of the waveform ($0 < t < 100$ ps), but in the last half ($t > 100$ ps), the waveform is dominated by the SRR-frequency component. Also, in the earliest times the signal amplitude are apparently saturated. Taking this into

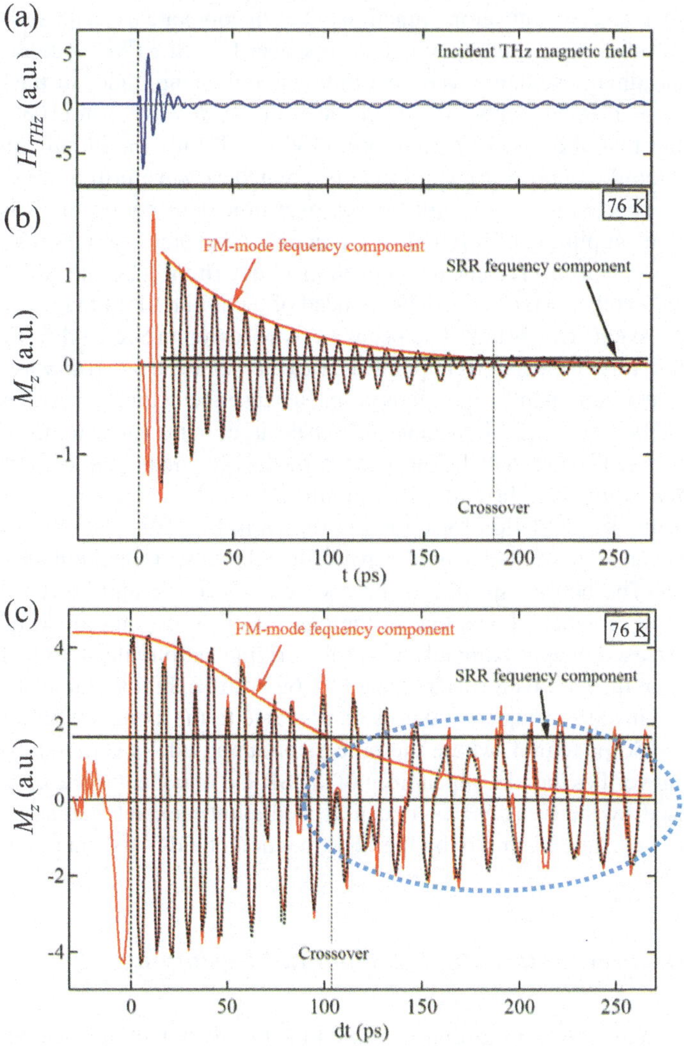

Fig. 5.8 **a** Temporal waveform of the incident THz magnetic field H_{THz} (t) used in calculation. **b** Calculated dynamics of z-component of magnetization after excitation by THz magnetic field (t-waveform) at 76 K. **c** Corresponding dt-waveform of (**b**). The black dotted curves in (**b**) and (**c**) are the fitting by Eqs. 5.1 and 5.2 in the main text

consideration, we fitted the dt-waveform with another function:

$$M_z^{final}(dt) = Erfc\left[\left\{A_1 \sin(\omega_1 dt - \phi_1)\exp\left(-\frac{dt - t_0}{\alpha_1}\right) + A_2 \sin(\omega_2 dt - \phi_2)\right\}/A_3\right] \quad (5.2)$$

We plotted again the envelope functions of each components as the red curve and black lines. Interestingly, the moment of crossover from the FM-mode-dominant to the SRR-dominant oscillation (i.e., the time when the amplitudes of the two envelope functions matches with each other) seen in the *dt*-waveform (105 ps) occurs much earlier than the case of *t*-waveform (180 ps). If only the initial amplitude of magnetization determines the amplitude of created *dt*-waveform at corresponding times, it is expected that such shift of crossover time does not occur. This suggests that when the amplitude of original spin precession becomes smaller than a certain level, the direction of created magnetization is determined by the SRR magnetic fields applied at the instance of SRPT, instead of the initial tilt of magnetization due to spin precession. This is the same behavior as observed in Sect. 4.4.6 [Fig. 4.21b].

In order to further verify this hypothesis, we calculated the *t*- and *dt*-waveforms for a lower temperature than the previous case. Because the detuning between the FM-mode and SRR resonance frequencies becomes larger in the lower temperatures, the precession signal is dominated almost solely by the FM-mode frequency component and the SRR-component becomes less significant.

The results of calculation for 69 K are shown in Fig. 5.9a. The SRR-frequency component cannot be visible in the t-waveform in this case throughout the calculated time region. The Fourier spectra of the *t*-waveforms for 69 and 76 K is plotted in Fig. 5.9b and c, which shows that in the case of 69 K the amplitude of the SRR peak is decreased to approximately 1/3 of the original amplitude at 76 K. In spite of this, however, the *dt*-waveform calculated for 69 K still exhibits a large amplitude of oscillation with SRR-frequency that is comparable to the signal amplitude at 76 K in the $dt > 150$ ps region, where the precession amplitude is sufficiently decayed. This finding verifies that the appearance of the SRR-component in the *dt*-waveforms is not caused by the weak SRR-frequency oscillation remnant in the original spin precession, but caused actually by the SRR magnetic fields surviving at the time.

5.2.2 Spin Dynamics During Potential Reshaping

The results in the previous sections showed that the creation of macroscopic magnetization due to THz magnetic *field* similar to what we observed in the experiment can be actually reproduced by using our free energy model, however, its physical picture remains unclear. In order to clarify the mechanism, we calculated the shape of transient free energy curves at each time and the relative direction of spins within it.

Figure 5.10a shows *t*-waveforms calculated for $T = 76$ K. Here, the heating pulse was incident on the sample at $t = 228$ ps, at which time the effect of SRR magnetic field, not the amplitude of precession, is expected to be the dominant cause of the symmetry breaking, as indicated from the result of previous subsetion. We calculate the free energy curves of the spin system projected onto x–z plane using the Eq. 2.1. The results are shown in Fig. 5.10(b), along with the circles indicating the direction of spin M_z at each time in x–z plane.

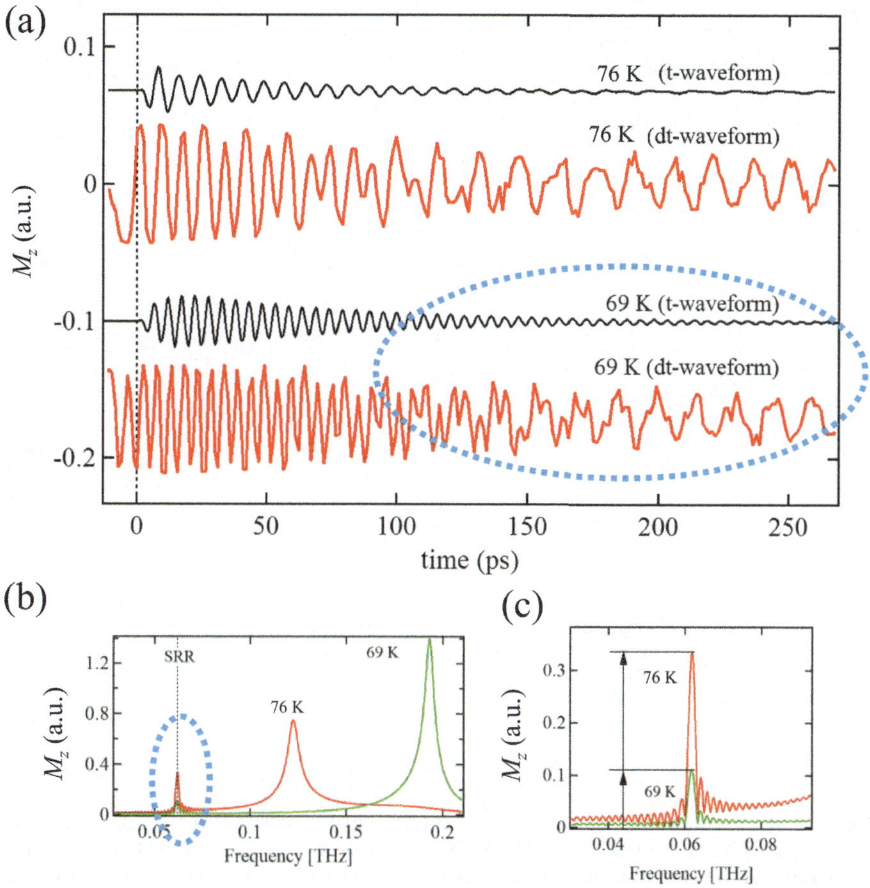

Fig. 5.9 **a** The *t*- and *dt*-waveforms calculated for 76 and 69 K. The region indicated by blue dotted circle is where the SRR-frequency component is dominant in the *dt*-waveform. **b** The Fourier spectra of the *t*-waveforms for 79 and 69 K. **c** Fourier spectra shown in (**b**) expanded around the SRR peak

Before the incidence of the THz magnetic field, magnetization M_z is fixed at the original position (*a*-axis (−60 ps)). During the action of impulsive THz magnetic field caused by higher mode of SRR, the strong magnetic field causes the shift of free energy curve towards either $-c$ (2.4 ps) or $+ c$ (4.8 ps) direction. This initiates the spin precession after the impulsive field of THz has gone (7.8 ps). At sufficiently later periods M_z remains almost still around the origin, with a tiny oscillation caused by the long-lived mode of SRR (222.6 ps).

Here, the ultrafast heating induced by the application of fs laser (228 ps) causes the free energy shape to change with time: the curvature of the parabola gradually opens up (235.2 ps), surpasses the state with almost a zero-curvature (236.4 ps) and immediately inverts its sign (240.6 ps). Note that M_z does not immediately follow the potential minima at each time, but remains for a while near the original position,

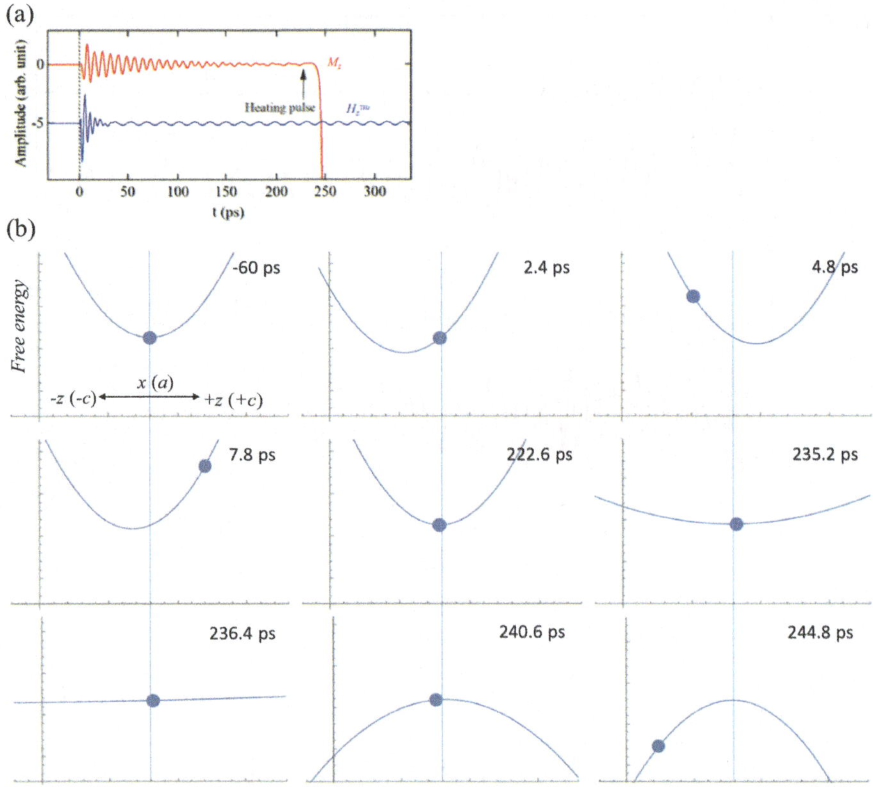

Fig. 5.10 **a** The t-waveforms calculated for 76 K (red) and the incident THz magnetic field (blue). The heating pulse is incident at t = 228 ps. **b** The free energy curves projected onto x–z plane calculated at each time. Horizontal axis indicates the angle [a-axis = π/2 (vertical line)]. The filled circle indicates the direction of M_z

which is an energetically unstable state after the shape of the parabola was changed to upward-convex. Once such state was created M_z starts to roll down the parabola (244.8 ps) towards the direction it was at the moment the upward-convex parabola was formed, which is to the left in this case. Finally, M_z settles into the new potential minima (−c-axis) in ~20 ps.

An important aspect can be foreseen from above result: the path of SRPT is ultimately determined by the relative position of M_z against the potential maxima within a short time after the change of sign from downward- to upward-convex. To further examine this process, we have expanded the above graphs around the minima and focused on the temporal region near the moment of this transition from downward- to upward-convex (Fig. 5.11b). Here, the position of potential minima (red) and maxima (black) are also plotted. Their dynamics are plotted as functions of time in Fig. 5.11a.

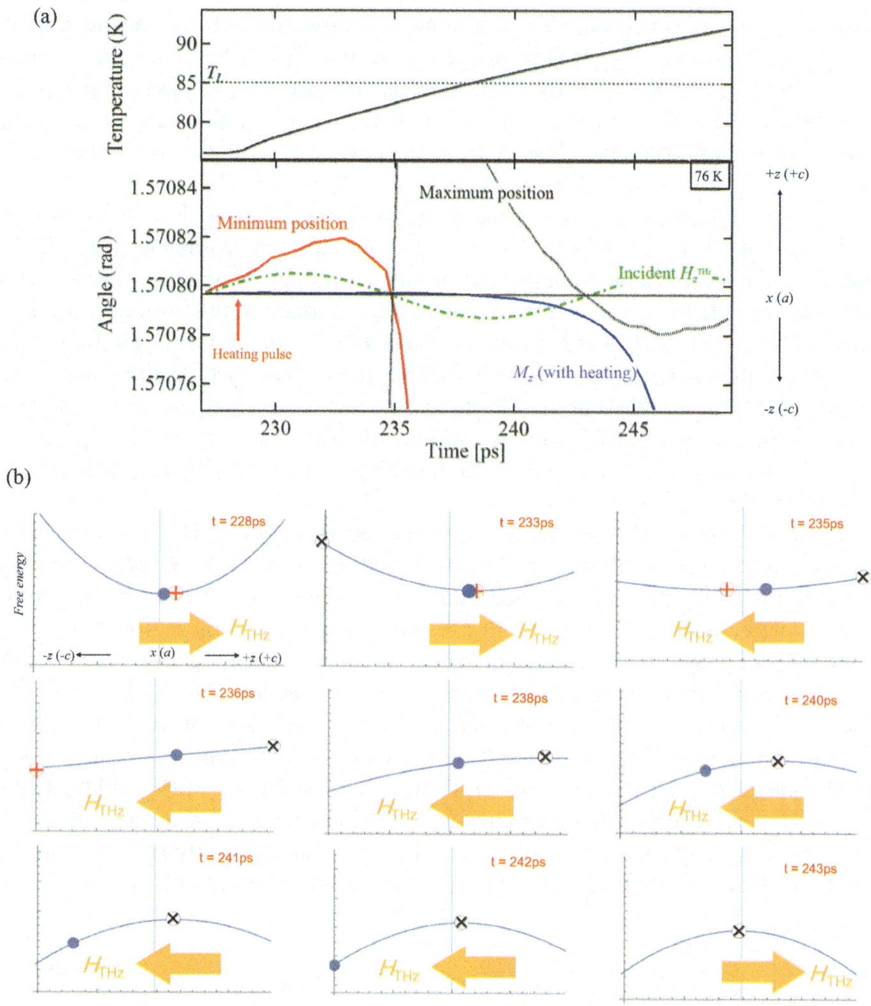

Fig. 5.11 **a** Simulated positions of M_z, angle of free energy minima (red) and maxima (black dashed), and incident THz magnetic field (green dotted) plotted as a function of time. The corresponding temperature is also indicated on the top graph. **b** The free energy curves, positions of M_z (blue circle), free energy minima (red vertical cross) and maxima (black cross) around the time of transition from downward- to upward-convex parabola. Directions of magnetic fields are also indicated

At the moment of this transition (t = 236 ps), it can be seen that due to the SRR magnetic field present at the time the flat potential is tilted slightly towards the left direction, while the position of M_z remains almost unchanged. Due to this, the potential minima is at far-left end and the maxima is at the rightmost side of the graph. As time progresses the curvature of the parabola increases (238–240 ps), making the point of potential maxima to gradually move leftward toward the origin. However, during this time the position of M_z is also pushed leftward due to the magnetic field, leaving away from the original *a*-axis (241–242 ps). Therefore, when the maxima had reached the origin, M_z is already at some distance from the *a*-axis (243 ps). After this, M_z cannot come back to the origin anymore because of the energy barrier and just keeps rolling down the slope until it reaches the new minimum located at –*c* axis.

The spin dynamics revealed in the above simulation can be directly applied to understand the behavior of an ensemble of spins in a similar manner (Fig. 5.12): during SRPT, there is a moment that the ensemble of spins remain in an energetically unstable state. If there was a THz magnetic field applied at this instance, the free energy curve is shifted to either direction and thus, spins are left in the middle of the slope at the moment of transition: therefore, the system starts "rolling down" the slope towards the direction of the THz field. As long as the THz field lasts for longer than the time it takes for M_z to move sufficiently away from the original position, M_z keeps moving in the same direction, resulting in the final states aligned in + *c*- or –*c*-directions.

A simplified overview of the abovementioned process of THz magnetic *field*-induced magnetization alignment is summarized in Fig. 5.12. When $H_{\mathrm{THz}} = 0$ (a), the maxima of the potential formed immediately after heating is at the origin (a-axis), which coincides the direction of the equilibrium direction of spin ensemble. As time progresses, the magnetic moments at each sides of the free-energy peak rotates towards the directions they were initially pointed at. This only leaves the final state with equal amount of +*c* and –*c* domains, unless there was a population shift towards either direction caused by the precession motion in advance to the irradiation of heating pulse. However, if $H_{\mathrm{THz}} \neq 0$ as shown in (b), although the equilibrium direction of the magnetization itself remains fixed near the original *a*-axis, the free-energy maxima are formed in a point that is slightly shifted from the origin by the magnetic field. From Eq. (2.1), the amount of this shift is estimated to

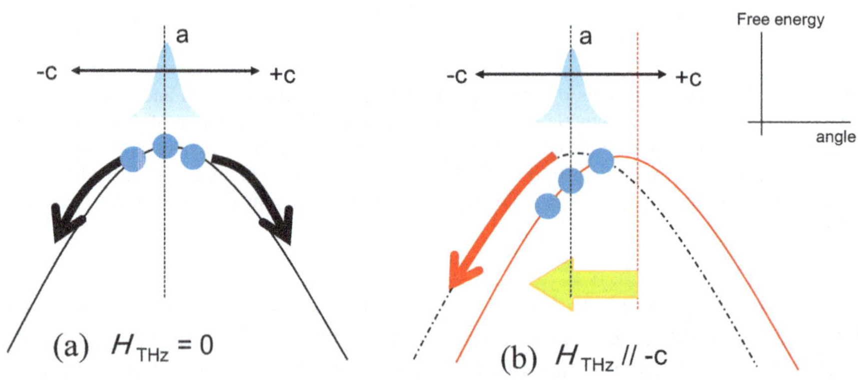

Fig. 5.12 Schematic of the mechanism of macroscopic domain alignment by THz magnetic fields. **a** The case of $H_{\mathrm{THz}} = 0$, and **b** $H_{\mathrm{THz}} \neq 0$

be ~ H_{THz} sinβ/4cos2β(A_{xx}-A_{zz}). Due to this shift of free energy curve relative to the average position of spin ensemble, the "division points" of the spin ensemble (NOT the population weight) upon SRPT is shifted and thus, larger portion of the spins rotates towards the direction of magnetic field.

As mentioned previously in Sect. 4.4.6, the mechanism revealed here cannot be verified if only the static magnetic field was used. Application of static magnetic field shifts the free energy maxima at the time of phase transition for sure, however, it also causes the initial equilibrium direction of spin ensemble to shift into the same direction. Therefore, the static magnetic field cannot create the condition wherein the magnetization is fixed at the origin and only the potential is shifted relative to it, as is readily achievable using the alternating magnetic fields of THz metamaterials.

5.3 Chapter Summary

In this chapter, we presented a numerical calculation of the spin dynamics based on a standard model using free energy of 2-spin system along with the LLG equation, in order to resolve the question raised from the experiment conducted in the previous chapter as to why the application of THz magnetic field can create macroscopically aligned magnetization states in the SRPT process without causing the population shift of the spin ensemble.

Through calculation with realistic parameters, we showed that most of the behaviors observed in the experiments, including the abovementioned phenomenon, can be understood within the framework of this model. Close examination of the transient free-energy curves and the time evolution of the magnetization indicated that instantaneous shift of the free energy curve relative to the original equilibrium direction of magnetization caused by the THz magnetic field is essential. Due to this change of free energy curve the division point for spin ensemble upon SRPT is shifted with respect to the original direction, resulting in the modulation of the portion of magnetic moments that rotate into the opposite final states. Such mechanism cannot be revealed if only the static magnetic field was used, and was realized for the first time owing to the usage of ultrafast alternating THz magnetic field produced by SRR.

It is worth mentioning that the time required for the determination of SRPT direction in the above process does not depend on the frequency of spin precession before irradiation of heating pulse. This indicates that it may be possible to generate macroscopically oriented magnetization states by using THz magnetic fields within a timescale much shorter than the precession itself, as long as the reshaping of free energy curve due to heating occurs fast enough.

References

1. Wood DL, Remeika JP, Kolb ED (1970) Optical Spectra of Rare-Earth Orthoferrites. J Appl Phys 41:5315
2. Yamaguchi K, Kurihara T, Minami Y, Nakajima M, Suemoto T (2013) Terahertz time-domain observation of spin reorientation in orthoferrite $ErFeO_3$ through magnetic free induction decay. Phys Rev Lett 110:137204

Chapter 6
Conclusion

Abstract This chapter summarizes the experimental and theoretical results obtained from previous chapters, followed by future remarks.

Keywords Ultrafast spintronics · Nearfield · Cavity QED · Intense THz fields

6.1 Summary

In this Thesis, we investigated the dynamical behavior of the magnetic order in erbium orthoferrite $ErFeO_3$ induced by the THz magnetic fields enhanced by a metallic structure. As mentioned in the Introduction part, ultrafast and nonthermal properties of the THz magnetic field have been so far successfully applied to probe magnetic dynamics of various materials. In previous studies, however, perturbation caused by THz magnetic fields applied on the spin systems have been limited to cause only tiny fluctuations, in contrast to the usage of THz electric fields which have been shown to be capable of not only probing but also causing a drastic change in the electronic states of condensed matter system.

In order to realize *macroscopic* THz-magnetic control of the spin order, we focused on a magnetic field-sensitive ultrafast process that takes place in $ErFeO_3$, i.e., laser-induced spin reorientation phase transition (SRPT). Furthermore, we introduced an intense THz light source along with the near-field enhancement effect in subwavelength metallic structure called split-ring resonator (SRR). By combining these concepts, we showed that:

(i) The SRR structure is actually capable of enhancing the spin precession amplitude by nearly an order of magnitude through resonant coupling.
(ii) Using such enhanced nearfield, it is possible to control the macroscopic magnetic order by applying it on the initial process of SRPT in $ErFeO_3$.

In Chap. 3 the abovementioned result (i) was obtained. Here we investigated the basic characteristic of the spin dynamics excited by the THz magnetic fields in SRR. We fabricated on the $ErFeO_3$ single crystal surface the metallic SRR structure whose frequency was precisely designed to match that of the spin precession of the $ErFeO_3$ substrate. By monitoring the transient Faraday rotation signals in

T. Kurihara, *Observation and Control of Magnetic Order Dynamics by Terahertz Magnetic Nearfield*, Springer Theses, https://doi.org/10.1007/978-981-16-8793-8_6

the THz pump-optical probe regime, it has been clearly observed that the amplitude of precession dynamics can be enhanced by approximately 8 times by using the resonant SRR fields, when the resonance frequencies of SRR and FM-mode spin precession were matched with each other. Also, Comparison between the spin precession dynamics with/without the SRR structure revealed that the lifetime of spin precession was shortened by the presence of SRR structure. Through numerical calculation using the Landau-Lifshitz-Gilbert (LLG) equation, it was revealed that there exists a mechanism of interactive energy transfer from the spin system to the SRR resonance mediated by magnetic near field, which has a close resemblance to the so-called Purcell effect well known in the study fields such as cavity quantum electrodynamics (QED).

In the experiment performed in Chap. 4 the abovementioned subject (ii) was studied. Here we proposed an experimental scheme to coherently control the macroscopic magnetic order using THz magnetic fields in the laser-induced SRPT process. We irradiated the $ErFeO_3$ sample using two different pump pulses of the optical heating pulse and the intense THz magnetic field enhanced by an SRR. Upon optical heat-induced SRPT, macroscopic magnetization rotates by 90° from $\boldsymbol{M}$ // a to $\boldsymbol{M}$ // $\pm c$. Under no external magnetic fields, the probability of rotation towards these two opposite states are equal and the optical heating leaves no net magnetization. However, by inducing the spin precession using THz magnetic field in advance to the heating pulse, the symmetry of these two transition paths are broken and only one of the two final states could be selected.

By changing the arrival times of heating pulse with respect to the THz pulse, it was clearly shown that the direction of magnetization created after SRPT was coherently inverted, indicating that the THz-induced spin precession can actually select the otherwise degenerated paths of SRPT towards up or down magnetization states through coherent spin precession. Unlike the previous researches in which only one optical pump pulse was used to simultaneously cause the heating and precession, our scheme of using two separate pumps enabled us to clearly distinguish these two effects. The result suggests that the coherent spin precession is crucial in the formation of macroscopic magnetization through SRPT.

By comparing the directions of the created magnetization inside/outside of the SRR structure, we have clearly shown that such process is actually induced by the pure *magnetic* THz nearfield of the SRR. At the two positions whose lateral distance is approximately only 1/16 of the incident THz spot size, the resulted magnetization direction was completely reversed. This implies that the usage of plasmonic structures enables the spatial resolution of THz magnetization-control to greatly exceed the diffraction limit, which is an especially important feature if similar techniques are to be implemented for device applications in the future.

It was indicated through THz amplitude-dependence measurement that in proper conditions, the created magnetization can be almost saturated uniformly in the heated area. Detailed analysis of the temperature-dependence measurement showed that the amount of created magnetization in the final state is determined by the balance between the instantaneous tilt of spins during precession, distribution width of the spin ensemble and the static tilt around the SRPT temperature region.

Furthermore, close comparison of the original spin precession dynamics with the time profile of the created magnetization in the final state showed that even when the spin precession amplitude was sufficiently small, a comparably large magnetization can be formed in the final state if THz magnetic field was simultaneously present. This counterintuitive result cannot be explained within the previous framework where only the initial *tilt of magnetization* from original axis selects the path of SRPT, and suggests another type of mechanism determined by the magnetic *field* itself applied at the moment of SRPT.

In order to clarify the mechanism of the observed macroscopic magnetization formation process under THz magnetic fields, in Chap. 5 we performed theoretical calculation of the spin dynamics based on LLG equation and standard 2-spin free energy model. Here we showed that basic features observed in the experiment can be qualitatively reproduced by the present model. By examining the temporal evolution of the free energy curves after optical heating, it was revealed that the transient THz magnetic field applied at the moment the sample temperature has reached the lower boundary of SRPT region T_L causes the shift of free energy maxima relative to the average position of the spin ensemble, which leads to the modification of the portion of spins to be rotated into opposite states upon subsequent SRPT and result in the alignment of magnetization towards either direction. It was shown that when the spin precession amplitude is small, this magnetic *field*-induced mechanism becomes the dominant cause of macroscopic magnetization formation. On the contrary, when precession is large enough, the conventional mechanism of magnetization tilting was shown to become dominant.

These calculation results showed that the instantaneous THz magnetic *field* itself actually is capable of creating a macroscopically aligned magnetization when properly applied during SRPT process. As the speed of such process is independent of the spin precession frequency before heating, this raises the possibility that as long as the heating process occurs fast enough, by using the THz magnetic fields it may be possible to create macroscopically magnetized state even within the duration period of the incident THz pulse, which is typically around 1 ps or shorter.

6.2 Future Remarks

Interesting applications can be foreseen in the usage of nearfield-coupling of the spin resonance and metamaterials revealed in Chap. 3. Magnetooptical effects are currently applied in the polarization devices such as isolator and circulator in the visible and microwave regions. Some materials (such as $SrFe_{12}O_{19}$ [1] or ε-Fe_2O_3 [2]) are reported to be suitable for such devices in the THz frequency region as well. On the other hand, in some of the recently invented metamaterial techniques, efficient polarization-control devices (giant optical activity in chiral structure [3], for example) have been realized using non-magnetic materials. By combining such artificial metamaterials with the THz spin systems, novel devices and electromagnetic phenomenon such as ultra-compact THz isolator and circulator, "optical diode" [4], frequency-tunable electromagnetically induced transparency (EIT) [5] and direct interconversion of magnon to THz photon [6], etc. may be realized, which would greatly extend the scope of THz-polarization devices in the Future. In order to realize such devices it is mandatory that we further understand the coupled behavior of the spin- and structure resonances, to which we gave the first insight in the present Thesis.

As for the THz-control of macroscopic magnetization state, the advantage of using the THz magnetic field for the spin control compared to other optical methods in the visible regions is that due to the small photon energy, only the spin systems can be resonantly excited without causing thermal effects or unwanted electronic transitions. Although our initial motivation of THz-control of macroscopic magnetization state has been achieved in Chap. 4, strictly speaking, this was assisted by optical heating and was not the *pure* magnetic control of the spin system. In that sense, it is more favorable to realize *all-THz* control of the macroscopic state of spin system using higher THz field strengths.

Very recently, there has been rapid developments in the ultra-strong THz light sources. For example, the quasi-CW THz source newly developed in Ref. [7] that exhibits peak THz amplitude of ~500 kV/cm and duration of several nanoseconds is suited for effectively exciting sharp resonant lines of elementary excitations in the THz region such as spin resonance studied in this Thesis. As another example of ultra-strong THz light source, the quasi half-cycle THz pulse that reaches peak electric field amplitude of ~80 MV/cm has been reported using the optical rectification in organic crystals [8, 9]. Combining such intense THz sources with the method of SRR-enhancement realized in Chap. 3 may enable the spin precession amplitude to reach an even higher level of nonlinear regime and, ultimately, magnetization reversal. To repeat, in order to avoid unwanted electronic excitations, it is mandatory to selectively enhance only the magnetic field of THz waves. The nearfield techniques such as proposed in the present Thesis are the best suited solution for this purpose.

The spin precession can be viewed as the propagating wave of angular momentum, or spin currents [10]. While the present operation speed of the spintronics devices are limited to around 10 GHz by the conventional electronics technology, usage of the THz fields will greatly increase this frequency limit to several hundreds to thousands of GHz. The abovementioned nonlinear dynamics of magnetization induced by the *all-THz* magnetic fields is expected to realize an unprecedented, completely non-thermal ultrafast spintronics devices based on the interconversion between spin- and electronic current in future, such as magnetic memory devices [11], THz detector based on inverse spin-Hall effects [12, 13], THz nano spin torque oscillator [14] and ultrafast spin transistor [15], etc. Although the diffraction limit may restrict the spatial resolution in such devices, the nearfield/metamaterial techniques have the potential to break this barrier by orders of magnitude.

References

1. Shalaby M, Peccianti M, Ozturk Y, Morandotti R (2013) A magnetic non-reciprocal isolator for broadband terahertz operation. Nat Commun 4:1558
2. Nakajima M, Namai A, Ohkoshi S, Suemoto T (2010) Ultrafast time domain demonstration of bulk magnetization precession at zero magnetic field ferromagnetic resonance induced by terahertz magnetic field. Opt Express 18:18260
3. Kuwata-Gonokami M, Saito N, Ino Y, Kauranen M, Jefimovs K, Vallius T, Turunen J, Svirko Y (2005) Giant optical activity in quasi-two-dimensional planar nanostructures. Phys Rev Lett 95:227401
4. Tomita S, Sawada K, Porokhnyuk A, Ueda T (2014) Direct observation of magnetochiral effects through a single metamolecule in microwave regions. Phys Rev Lett 113:235501
5. Papasimakis N, Fedotov VA, Zheludev NI, Prosvirnin SL (2008) Metamaterial Analog of Electromagnetically Induced Transparency. Phys Rev Lett 101:253903
6. Tabuchi Y, Ishino S, Noguchi A, Ishikawa T, Yamazaki R, Usami K, Nakamura Y (2015) Coherent coupling between a ferromagnetic magnon and a superconducting qubit. Science 349:405
7. Hayashi S, Nawata K, Taira T, Shikata J, Kawase K, Minamide H (2015) Ultrabright continuously tunable terahertz-wave generation at room temperature. Sci Rep 4:5045
8. Vicario C, Monoszlai B, Hauri CP (2014) GV/m single-cycle terahertz fields from a laser-driven large-size partitioned organic crystal. Phys Rev Lett 112:213901
9. Shalaby M, Hauri CP (2015) Demonstration of a low-frequency three-dimensional terahertz bullet with extreme brightness. Nat Commun 6:5976
10. Wolf SA, Awschalom DD, Buhrman RA, Daughton JM, Molnár S, Roukes ML, Chtchelkanova AY, Treger DM (2001) Spintronics: a spin-based electronics vision for the future. Science 294(5546):1488
11. Zhou N, Xu X, Hammack AT, Stipe BC, Gao K, Scholz W, Gage EC (2014) Plasmonic near-field transducer for heat-assisted magnetic recording. Nanophotonics 3:141
12. Kimura T, Otani Y, Sato T, Takahashi S, Maekawa S (2007) Room-temperature reversible spin hall effect. Phys Rev Lett 98:156601
13. Saitoh E, Ueda M, Miyajima H, Tatara G (2006) Conversion of spin current into charge current at room temperature: inverse spin-hall effect. Appl Phys Lett 88:182509
14. Kiselev SI, Sankey JC, Krivorotov IN, Emley NC, Schoelkopf RJ, Buhrman RA, Ralph DC (2003) Microwave oscillations of a nanomagnet driven by a spin-polarized current. Nature 425:380
15. Datta S, Das B (1990) Electronic analog of the electro-optic modulator. Appl Phys Lett 56:665

Appendix
Relation Between the *s*- and *m*-parameters

In Chap. 3 we introduced the interactive coupling between SRR and spin resonances in terms of independent parameters s and m, corresponding to the coefficients of spin excitation by SRR magnetic field and the SRR excitation by spin precession due to electromotive force, respectively. In this section, we analytically derive the relation between these coefficients.

The procedure is as follows: first, we write down the Hamiltonian that describes non-interacting systems including SRR and spin. Next, we add an interaction term between them in the Hamiltonian. Lastly, the equations of motion for both systems are derived using the new Hamiltonian. The interaction term of Hamiltonian gives rise to the driving force-term in each systems. By comparing these terms with s- and m-parameters, the analytical relation between them can be deduced.

A.1 Hamiltonian of SRR and Spin Systems Without Interaction

We start from writing down the Hamiltonian that describes the SRR and spin systems. For simplicity, we neglect damping terms ($\mathrm{R} = 0$, $\alpha = 0$). The external forces are also neglected ($\boldsymbol{H}_{\mathrm{THz}} = 0$, $\boldsymbol{E}_{\mathrm{THz}} = 0$). Then, Eqs. (3.1 to (3.4) can be deduced into the following sets of equations:

$$L\frac{d^2 I(t)}{dt^2} + \frac{1}{C} I(t) = m\frac{d^2 M_{1z}(t)}{dt^2} \tag{A.1}$$

$$\frac{d\boldsymbol{M}}{dt} = -\gamma \boldsymbol{M} \times \boldsymbol{H}_{eff} \tag{A.2}$$

$$\boldsymbol{H}_{eff}(t) = H_0\hat{\boldsymbol{x}} + sLI(t)\hat{\boldsymbol{z}} \tag{A.3}$$

T. Kurihara, *Observation and Control of Magnetic Order Dynamics by Terahertz Magnetic Nearfield*, Springer Theses, https://doi.org/10.1007/978-981-16-8793-8

If we neglect the interaction terms ($s = m = 0$), they reduce to:

$$\begin{cases} L\frac{d^2 I(t)}{dt^2} + \frac{1}{C} I(t) = 0 \\ \frac{dM}{dt} = -\gamma M \times \boldsymbol{H}_0 \end{cases} \tag{A.4}$$

where $\boldsymbol{H}_0 = H_0\, \hat{\boldsymbol{x}}$. The Hamiltonian of each system can be expressed as follows:

$$\mathcal{H}_{SRR} = \frac{Q^2}{2C} + \frac{LI^2}{2} \tag{A.5}$$

$$\mathcal{H}_{spin} = -\gamma \boldsymbol{M} \cdot \boldsymbol{H}_0 \tag{A.6}$$

where Q and I are the charge in the capacitor and current in the coil, respectively.

Equation A.5 can be rewritten in terms of conjugate parameters Q (conjugate coordinate) and P (conjugate momenta) that satisfy Hamilton's relations:

$$\dot{Q} = +\frac{\partial \mathcal{H}_{SRR}}{\partial P} \equiv I \tag{A.7}$$

$$\dot{P} = -\frac{\partial H_{SRR}}{\partial Q} = -\frac{Q}{C} \tag{A.8}$$

In order for these equations to stand, $P = LI$ must stand. Therefore, eq. A.5 becomes:

$$\mathcal{H}_{SRR} = \frac{Q^2}{2C} + \frac{P^2}{2L} \tag{A.9}$$

$$\mathcal{H}_{spin} = -\gamma \boldsymbol{M} \cdot \boldsymbol{H}_0 \tag{A.10}$$

It should be noticed that Q and P are connected with each other through the following commutation relation:

$$[Q, P] = i \tag{A.11}$$

Using A.11, from Heisenberg equation of motion we can reproduce the original equation of motion:

$$\dot{Q} = -i[Q, \mathcal{H}_{SRR}] = -i\left[Q, \frac{P^2}{2L}\right] = \frac{\mathrm{P}}{\mathrm{L}} \tag{A.12}$$

$$\therefore \ddot{Q} = \frac{\dot{P}}{L} = -\frac{i[P, \mathcal{H}_{SRR}]}{L} = -\frac{i\left[P, \frac{Q^2}{2C}\right]}{L} = -\frac{Q}{LC} \tag{A.13}$$

which is equivalent to eq. A.3. In a similar manner, the equation of motion for the spin system can be reproduced from eq. A.10 using Heisenberg equation and commutation relations.

As the total Hamiltonian of the system, so far we get:

$$\mathcal{H}_{tot} = \mathcal{H}_{SRR} + \mathcal{H}_{spin} = \frac{Q^2}{2C} + \frac{P^2}{2L} - \gamma \boldsymbol{M} \cdot \boldsymbol{H}_0 \tag{A.14}$$

where $P = LI$.

A.2 Inclusion of Interaction Term into Total Hamiltonian

As the second step, we include in the total Hamiltonian eq. A.14 a term that describes interactive energy transfer between them. As the simplest form of interaction, we can suppose:

$$\mathcal{H}_{int} = -kM_z P = -kM_z LI, \tag{A.15}$$

which means that the magnetic field produced by the current I inductively interacts with the magnetic dipole moment M_z of the spin system. The total Hamiltonian including the above interaction then becomes:

$$\mathcal{H}_{tot} = \mathcal{H}_{SRR} + \mathcal{H}_{spin} + \mathcal{H}_{int} = \frac{Q^2}{2C} + \frac{P^2}{2L} - \gamma \boldsymbol{M} \cdot \boldsymbol{H}_0 - kM_z P \tag{A.16}$$

A.3 Retrieving *s* and *m* from the New Hamiltonian

Next, we reproduce the equations of motion for both the SRR and spin systems from the new Hamiltonian A.16 using Heisenberg equation and commutation relations.

(i) *Equation of Motion of the SRR System*

Equation of motion for Q is given as:

$$\dot{Q} = -i[Q, \mathcal{H}_{tot}] = -i\left[Q, \frac{P^2}{2L}\right] - i[Q, kM_z P] = \frac{P}{L} + kM_z \tag{A.17}$$

$$\therefore P = L(\dot{Q} - kM_z) = L(I - kM_z) \tag{A.18}$$

On the other hand, for P:

$$\dot{P} = -i[P, \mathcal{H}_{tot}] = -i\left[P, \frac{Q^2}{2C}\right] - \frac{i}{2C}[P, Q^2] = -\frac{Q}{C} \tag{A.19}$$

Here, from A.18,

$$\dot{P} = L(\ddot{Q} - k\dot{M}_z) \tag{A.20}$$

Combining this with eq. A.19, we can get:

$$\begin{aligned} L\ddot{Q} + \frac{Q}{C} - Lk\dot{M}_z = 0 \\ L\dddot{Q} + \frac{\dot{Q}}{C} - Lk\ddot{M}_z = 0 \\ \therefore L\ddot{I} + \frac{I}{C} - Lk\ddot{M}_z = 0 \end{aligned} \tag{A.21}$$

which is equivalent to the original equation of motion eq. A.1, if $m = kL$.

(ii) *Equation of Motion of the Spin System*

The commutation relations for spin are $[M_x, M_y] = iM_z$, $[M_y, M_z] = iM_x$, and $[M_z, M_x] = iM_y$. Using this,

$$\begin{aligned} \dot{M}_x &= -i[M_x, \mathcal{H}_{tot}] = -\gamma(\boldsymbol{M} \times \boldsymbol{H}_0)_x + i[M_x, kM_zP] \\ &= -\gamma(\boldsymbol{M} \times \boldsymbol{H}_0)_x + kM_yP = -\gamma\left\{\boldsymbol{M} \times \left[\boldsymbol{H}_0 + \begin{pmatrix} 0 \\ 0 \\ kP \end{pmatrix}\right]\right\}_x \end{aligned} \tag{A.22}$$

$$\begin{aligned} \dot{M}_y &= -i[M_y, \mathcal{H}_{tot}] = -\gamma(\boldsymbol{M} \times \boldsymbol{H}_0)_y + i[M_y, kM_zP] \\ &= -\gamma(\boldsymbol{M} \times \boldsymbol{H}_0)_y - kM_xP = -\gamma\left\{\boldsymbol{M} \times \left[\boldsymbol{H}_0 + \begin{pmatrix} 0 \\ 0 \\ kP \end{pmatrix}\right]\right\}_y \end{aligned} \tag{A.23}$$

$$\dot{M}_z = -i[M_z, \mathcal{H}_{tot}] = -\gamma(\boldsymbol{M} \times \boldsymbol{H}_0)_z = -\gamma\left\{\boldsymbol{M} \times \left[\boldsymbol{H}_0 + \begin{pmatrix} 0 \\ 0 \\ kP \end{pmatrix}\right]\right\}_z \tag{A.24}$$

$$\therefore \dot{\boldsymbol{M}} = -\gamma \boldsymbol{M} \times \left(\boldsymbol{H}_0 + \frac{kP}{\gamma}\hat{z}\right) \tag{A.25}$$

It can be seen that the coupling term (A.15) previously introduced acts on the spin system as an effective magnetic field. Inserting eq. A.18 into A.25,

$$\therefore \dot{\boldsymbol{M}} = -\gamma \boldsymbol{M} \times \left(\boldsymbol{H}_0 + \frac{k}{\gamma}(I - kM_z)\hat{z} \right) \tag{A.26}$$

If we assume that the interaction is weak (k << 1), we can neglect the last term which is quadratic to k ($k^2 M_z/\gamma$). Then, eq. A.26 is equivalent to the original equation of motion A.2, A.3, where $s = k/\gamma$.

To summarize, coupling constant s and m can be derived from single parameter k from the following relations:

$$s = \frac{k}{\gamma} \tag{A.27}$$

$$m = kL \tag{A.28}$$

Thus, it was analytically proved that the coupling parameters m and s are not independent of each other but rather directly connected by the relation $s = m/L\gamma$.

A.4 Remark

The last term in eq. A.26 which we neglected is antiparallel to the magnetic moment M_z. It suggests that this term can be regarded as a kind of demagnetizing field. It is interesting to note that in conventional ferromagnetic resonance, demagnetizing field strongly modifies the trajectory of the spin precession from a circular into an elliptical one. During such elliptic motion, it is known that a strong second harmonic signal can be generated (SHG) along the direction of equilibrium magnetic moment. This suggests that the usage of metamaterials may enable the nonlinearity of spin dynamics to be enhanced if electromagnetic coupling was strong enough (k >> 1).

The manufacturer's authorised representative in the EU is Springer Nature Customer Service Centre GmbH, Europaplatz 3, 69115 Heidelberg, Germany. If you have any concerns regarding our products, please contact ProductSafety@springernature.com

Printed and bound by CPI Group (UK) Ltd, Croydon, CR0 4YY
30/09/2026
02225307-0003